Smart Agriculture

Volume 11

Series Editors

Zhao Zhang, Key Laboratory of Smart Agriculture System Integration Research, Ministry of Education, China Agricultural University, Beijing, China

Yiannis Ampatzidis, UF/IFAS Southwest Florida Research and Education Center, University of Florida, Immokalee, FL, USA

Paulo Flores, Department of Agricultural and Biosystems Engineering, North Dakota State University, Fargo, ND, USA

Yuanjie Wang, Agricultural Information Institute, Chinese Academy of Agricultural Sciences, Beijing, China

The book series Smart Agriculture presents progress of smart agricultural technologies, which includes, but not limited to, specialty crop harvest robotics, UAV technologies for row crops, innovative IoT applications in plant factories, and big data for optimizing production process. It includes both theoretical study and practical applications, with emphasis on systematic studies. AI technologies in agricultural productions will be emphasized, consisting of innovative algorithms and new application domains. Additionally, new crops are emerging, such as hemp in U.S., and covered as well. This book series would cover regions worldwide, such as U.S., Canada, China, Japan, Korea, and Brazil.

The book series Smart Agriculture aims to provide an academic platform for interdisciplinary researchers to provide their state-of-the-art technologies related to smart agriculture. Researchers of different academic backgrounds are encouraged to contribute to the book, such as agriculture engineers, breeders, horticulturist, agronomist, and plant pathologists. The series would target a very broad audience – all having a professional related to agriculture production. It also could be used as textbooks for graduate students.

Can Hu · Long Wang · Jianfei Xing ·
Xufeng Wang · Zhao Zhang
Editors

New Technologies Applied in Gobi Desert Facility Agriculture

Sensing and Autonomous Techniques

Springer

Editors
Can Hu
Tarim University
Aral, Xinjiang, China

Long Wang
Tarim University
Aral, Xinjiang, China

Jianfei Xing
Tarim University
Aral, Xinjiang, China

Xufeng Wang
Tarim University
Aral, Xinjiang, China

Zhao Zhang
Key Laboratory of Smart Agriculture
System Integration Research
China Agricultural University
Beijing, China

ISSN 2731-3476 ISSN 2731-3484 (electronic)
Smart Agriculture
ISBN 978-981-96-1030-3 ISBN 978-981-96-1031-0 (eBook)
https://doi.org/10.1007/978-981-96-1031-0

This Springer imprint is published by the registered company Springer Nature Singapore Pte Ltd.
The registered company address is: 152 Beach Road, #21-01/04 Gateway East, Singapore 189721, Singapore

Preface

To satisfy the increasing agricultural produce demands of expanding population, the Gobi desert, which has the harshest environment for agriculture and has long been considered unsuitable for agricultural production, has been recently been utilized to produce fresh fruits and vegetables. The Gobi desert facility agriculture, which is a relative new technology, has significantly transformed the Gobi desert. Innovative sensing and autonomous technique has recently been widely applied in Gobi desert facility agriculture. Though many papers, articles, theses, and dissertations have been published to describe the new technologies applied in Gobi desert facility agriculture production, there lacks a book to comprehensively summarize the progress.

To fill the gap of missing a book to systematically and comprehensively discuss the technology progress in Gobi desert facility agriculture, this book is proposed, which consists of 10 chapters. Chap. 1 generally presents current status of research on automated operations of Gobi desert facilities, followed by Chaps. 2 and 3 in cultivation model of yellow sand substrate and designing/testing of a self-propelled trenching and mulching machine for desert Gobi sand substrate, respectively. Then, Chaps. 4 and 5 depict the study on intelligent identification and transplanting technology of plug seedlings, designing intelligent transplantation device, respectively. Chapter 6 focuses on water and fertilizer demand model on fruit and vegetable production, and Chap. 7 illustrates intelligent water and fertilizer integration system, followed by Chaps. 8 and 9 on spraying agent model and intelligent spraying machine, respectively. The last but not least Chap. 10 is related to mechanized operation mode of the whole process of facility agriculture in Gobi desert.

Drs. Can Hu, Long Wang, Jianfei Xing, Xufeng Wang, and Zhao Zhang, the editors of this book titled New Technologies Applied in Gobi Desert Facility Agriculture—Sensing and Autonomous Techniques have been working in the area of Gobi desert facility agriculture for decades, and their expertise and knowledge make a sound base for the book's good quality. This book provides state-of-the-art information for readers who are interested in learning about the important subject of innovative sensing and autonomous systems applied in Gobi desert facility agriculture.

Aral, China Can Hu
Aral, China Long Wang
Aral, China Jianfei Xing
Aral, China Xufeng Wang
Beijing, China Zhao Zhang

Contents

Chapter 1
Current Status of Research on Automated Operations of Desert Gobi Facilities

Jiahui Xu, Jianfei Xing, Long Wang, Yalong Song, Xufeng Wang, and Zhaoyu Rui

1.1 Introduction

As the global population continues to grow and the impact of climate change becomes increasingly apparent, how to achieve sustainable agricultural production in extreme environments has become an important global challenge. The Gobi desert region,

J. Xu · J. Xing (✉) · L. Wang · Y. Song · X. Wang
College of Mechanical and Electrical Engineering, Tarim University, Alar 843300, China
e-mail: 591101353@qq.com

Modern Agricultural Engineering Key Laboratory at Universities of Education Department of Xinjiang Uygur Autonomous Region, Alar 843300, China

Xinjiang Production and Construction Corps (XPCC) Key Laboratory of Utilization and Equipment of Special Agricultural and Forestry Products in Southern Xinjiang, Alar 843300, China

J. Xu
e-mail: 1760749323@qq.com

L. Wang
e-mail: 120140002@taru.edu.cn

Y. Song
e-mail: 315327137@qq.com

X. Wang
e-mail: wxfwyq@126.com

Z. Rui
Key Laboratory of Smart Agriculture System Integration, Ministry of Education, Beijing 100083, China
e-mail: 1040160789@qq.com

Key Laboratory of Agricultural Information Acquisition Technology, Ministry of Agriculture and Rural Affairs, Beijing 100083, China

C. Hu et al. (eds.), *New Technologies Applied in Gobi Desert Facility Agriculture*, Smart Agriculture 11, https://doi.org/10.1007/978-981-96-1031-0_1

as one of the harshest environments on earth, has long been considered unsuitable for agricultural production. However, in recent years, with the rapid development of facility agriculture technology and the widespread application of automation systems, these areas that were once considered "barren lands" are gradually showing great agricultural potential.

Desert Gobi facility agriculture is an efficient agricultural production model carried out under a controlled environment. It uses greenhouses, plastic greenhouses and other facilities to create a microenvironment suitable for crop growth, effectively overcoming the poor soil, poor water retention, and severe salinization in desert Gobi areas. It uses the local abundant yellow sand resources as a cultivation matrix, which can not only solve soil problems, but also realize the recycling of resources. However, traditional facility agriculture still faces many challenges in desert Gobi areas, such as labor shortage, low resource utilization efficiency, and difficulty in environmental control. Against this background, it has become necessary to realize Gobi facility automation technology. The application of automation technology in desert Gobi facility agriculture can not only improve production efficiency and reduce labor intensity, but also realize precise management, optimize resource utilization, and improve crop yield and quality. From automatic irrigation systems, intelligent water and fertilizer integration, to agricultural robots, and Internet of Things monitoring platforms, facility automation technology is changing the production model of desert Gobi facility agriculture in all aspects. In recent years, with the rapid development of emerging technologies such as intelligence, big data, and the Internet of Things, the intelligence level of desert Gobi facility agriculture automation systems has been continuously improved. For example, the crop growth status monitoring system based on machine learning can adjust the growth environment parameters in real time; the intelligent irrigation system can automatically perform precise irrigation based on data such as soil moisture and crop water requirement; agricultural robots can complete multiple tasks such as sowing, fertilizing, and harvesting, greatly reducing manpower requirements.

We will first analyze the characteristics of the desert Gobi environment and its impact on the automated operation of facility agriculture, and then discuss in detail the main application areas and specific application status of automation technology in desert Gobi facility agriculture. Next, we will focus on the key technologies of current automated operations and conduct a comparative analysis of the current research status at home and abroad. On this basis, we will deeply explore the challenges faced by automated operations in desert Gobi facility agriculture and look forward to future development trends.

Through a comprehensive review and analysis of the current research status, we hope to provide valuable references for subsequent technological innovation and application practices. At the same time, we also hope to attract more researchers and practitioners to the automated operations of desert Gobi facility agriculture and promote the rapid development of this field. In the context of global food security and sustainable development, improving agricultural production efficiency and sustainability in desert Gobi areas has important practical significance. The widespread application of automation technology will undoubtedly provide strong technical support for achieving this goal.

This book will provide readers with a comprehensive and systematic perspective to understand the latest research progress, challenges and future development directions of automated operations in desert Gobi facility agriculture. Through these contents, we hope to inspire more innovative thinking and provide theoretical basis and practical guidance for sustainable agricultural development in desert areas and even in arid and semi-arid areas around the world through systematic exploration of these innovative technologies.

1.2 Desert Gobi Environmental Characteristics and Their Impact on Facility Agriculture Automation Operations

1.2.1 Climate Conditions

Understanding the desert Gobi environment is the basis for conducting research on facility agriculture automation operations. The desert Gobi region is famous for its extreme climatic conditions, which poses a huge challenge to the use of facility agriculture automation equipment. The temperature changes in the desert and Gobi regions are extreme, and the temperature difference between day and night can reach tens of degrees Celsius. The daytime temperature in summer often exceeds 40 °C, while the temperature at night in winter may drop to below zero. This extreme temperature difference places high demands on the material properties of the temperature control system and automation equipment in facility agriculture, and the equipment must have good heat resistance and cold resistance. The annual precipitation in the desert Gobi region is extremely small, usually less than 50 mm, and even less than 25 mm in some areas. This makes water resources extremely scarce, which is a huge challenge for agricultural production that relies on water resources. Automated irrigation systems can maximize the use of limited water resources and improve water resource utilization efficiency through precise control. Desert Gobi regions often encounter strong windy and sandy weather, with wind speeds of up to tens of kilometers per hour, which causes a large amount of sand and dust particles in the air. These sand and dust particles not only affect the growth of crops, but also bring additional burdens to the operation and maintenance of automation equipment. Mechanical equipment in facility agriculture needs to have a dustproof design to ensure its long-term reliable operation in this environment. The surface of desert and Gobi areas is usually uneven, with a large number of sand dunes and gravel, which poses a challenge to the construction of facility agriculture and the stable operation of automated equipment. Leveling the ground and building a solid agricultural infrastructure are essential preparations. The soil in desert areas is usually loose sandy and lacks organic matter and minerals. In facility agriculture, it is necessary to add fertilizers and other soil conditioners to improve the fertility and structural stability of the soil to ensure that crops can grow normally. Automated farming equipment needs to be designed to adapt to the working mode of soft soil to avoid sinking or damage.

The ecosystem in the desert Gobi region is relatively fragile, and any large-scale human activities are likely to cause irreversible damage to the local environment. Due to the harsh climatic conditions, the biodiversity in the desert Gobi region is low and the ecosystem stability is poor. Facility agriculture must consider how to produce efficiently while minimizing the damage to the ecological environment. The introduction of automated equipment can reduce the interference of human activities and reduce the negative impact on the ecological environment. The desert Gobi region itself is also facing the problem of land degradation and desertification. The application of facility agriculture and automation technology should focus on sustainable development and adopt technical measures to prevent and control desertification, such as vegetation restoration and soil improvement, to prevent or slow down the process of desertification.

In summary, the harshness and complexity of the desert Gobi environment, although it has brought various challenges, also provides a broad application scenario and research space for facility agriculture automation technology. Through a deep understanding of environmental characteristics, we can better design and apply automation systems that adapt to this special environment to improve the efficiency and sustainability of agricultural production.

1.2.2 Geographical Characteristics and Ecological Environment Conditions

The Gobi desert is one of the most extreme and unique environments on earth, and its geographical features and ecological environment show amazing adaptability and diversity. This vast desert area is known for its arid, hot climate, with annual rainfall typically not exceeding 250 mm, making water a scarce and precious resource. Topographically, the Gobi Desert is characterized by undulating sand dunes, rocky deserts and vast gravel plains, forming a unique and spectacular landscape. One of the characteristics of the climate here is the huge temperature difference, whether between day and night or between seasons, the temperature changes are extremely significant. The sun is scorching during the day and the temperature may reach over 40 °C, while at night it may plummet to close to freezing. Strong winds are another notable feature of the Gobi Desert. The perennially prevalent strong winds not only shape the unique wind-eroded landforms, but are also the main cause of the formation of sandstorms. Such harsh climatic conditions have created an extremely fragile ecosystem that is extremely sensitive to external interference. Once damaged, its ability to recover is extremely limited. Despite this, life still survives tenaciously in this seemingly barren land. Although the vegetation is sparse, it has shown amazing adaptability, such as cacti, cypress and other drought-tolerant and salt-tolerant plants.

They adapt to extreme environments through special physiological mechanisms. The animal kingdom has also shown powerful survival skills, such as camels that can go without water for long periods of time and gerbils that can obtain water from food. However, due to environmental constraints, the biodiversity of the desert Gobi is relatively low, and the number of species is far less than that of other ecosystems. The soil is poor, with low organic matter content and a serious lack of nutrients, which further limits the reproduction and growth of organisms. Despite this, oases are still formed in places with relatively abundant water sources, becoming green islands of life in the desert, providing habitat and water sources for many creatures. The temperature difference between day and night in the Gobi desert not only affects the survival of organisms, but also shapes unique landforms. Rocks continue to expand and contract under severe temperature changes, eventually forming unique weathered landforms. This landform not only has scientific research value, but has also become a natural wonder that attracts tourists. Despite its harsh environment, the Gobi desert still plays an important role in global ecosystems. They regulate global climate, influence atmospheric circulation, and even transport minerals to distant ecosystems through dust storms. In addition, it is also rich in mineral resources and has become an important economic development area. However, human activities are putting unprecedented pressure on this fragile ecosystem. Over-development, irrational water resource utilization, global climate change and other factors are aggravating the desertification process, threatening the already fragile ecological balance here. Therefore, how to find a balance between development and utilization and ecological protection has become a major challenge currently faced.

Overall, the Gobi desert environment is an extreme, unique and fragile ecosystem. It not only demonstrates the resilience and adaptability of life, but also provides us with valuable opportunities to study the evolution of life in extreme environments. Understanding and protecting this unique ecological environment is not only crucial to local sustainable development, but also has far-reaching significance for global ecological balance (Fig. 1.1).

1.3 The Necessity of Automated Operations in Desert Gobi Facilities

The desert Gobi environment has extremely unique and severe characteristics, which have a profound impact on automated agricultural operations. Understanding these environmental characteristics and their effects is critical to developing and optimizing automated agricultural equipment and systems suitable for the desert Gobi region. Extreme drought is the most striking feature of the Gobi Desert. Annual precipitation usually does not exceed 200 mm, and in some areas it is even less than 50 mm. This extremely water-scarce environment poses severe challenges to automated agricultural operations. The scarcity of water resources requires irrigation systems to be extremely precise and efficient. Therefore, automated irrigation equipment needs

Fig. 1.1 Facility agriculture automation greenhouse

to be equipped with advanced sensor networks that can monitor parameters such as soil moisture, crop water demand and evaporation in real time. At the same time, the irrigation system also needs to have intelligent decision-making capabilities and automatically adjust irrigation strategies based on these data to ensure that every drop of water is used most efficiently. In addition, water recovery and recycling systems are particularly important in this environment, and automated equipment that can efficiently treat and purify wastewater needs to be developed to maximize water resource utilization. The temperature difference between day and night in the Gobi Desert is extremely large, which places higher requirements on the greenhouse environmental control system. Daytime temperatures can reach over 40 °C, while nighttime temperatures can be close to freezing. This severe temperature fluctuation requires the automated temperature control system to have fast response and precise adjustment capabilities. The system needs to be equipped with multiple temperature sensors to monitor temperature changes at different locations in the greenhouse in real time, and can automatically adjust the operation of refrigeration or heating equipment according to the growth stage and variety characteristics of the crops. At the same time, considering energy consumption issues, the temperature control system also needs to have intelligent energy-saving strategies, such as using the temperature difference between day and night for passive cooling or heat storage. Strong sunlight and high ultraviolet radiation are another notable feature of the Gobi Desert. Although sufficient light is beneficial to crop photosynthesis, excessive sunlight and ultraviolet rays may also cause damage to crops. Therefore, automated shading systems have become particularly important. Such systems need to be able to automatically adjust the opening and closing of sunshades based on light intensity and crop growth needs. At the same time, considering the strong winds in desert areas, the shading system also needs to have sufficient wind resistance. Sandstorms are a major threat in the Gobi Desert. Frequent sandstorms not only cause direct damage to crops, but also affect the normal operation of equipment. This requires that all automation equipment must have good dust-proof performance. For example, sensors require special protection designs to avoid dust interference; mechanical equipment such as seeders and harvesters need to have self-cleaning functions to prevent dust accumulation from affecting their performance. In addition, special automated cleaning systems need to be developed to regularly remove sand and dust accumulation from equipment and greenhouse surfaces. The soil in the Gobi Desert is poor and has low organic matter content, which places special requirements on automated fertilization systems. The system needs to be able to accurately monitor soil nutrient status and automatically prepare the best fertilizer formula according to the needs of different crops. Taking into account environmental protection and resource conservation, the fertilization system should also have a recycling function to minimize nutrient loss. Relatively low biodiversity is another characteristic of the Gobi Desert, but it still faces specific threats from pests and diseases. Automated pest monitoring and control systems need to be optimized for common local pests and diseases. For example, developing image recognition systems that can early identify special pests such as desert locusts, and automatic spraying equipment that can effectively apply pesticides in extremely dry environments.

The extreme environmental characteristics of the Gobi desert pose severe challenges to automated agricultural operations, but at the same time it also promotes the innovation and development of related technologies. Automation systems adapted to these special environments not only need to have the characteristics of high precision and efficiency, but also need to have strong environmental adaptability and anti-interference capabilities. Through continuous optimization and innovation, the desert Gobi facility agricultural automation operating system will open up new possibilities for agricultural development in these extreme areas and make important contributions to solving global food security issues.

1.4 Current Status of Mechanization Development of Gobi Facility Agriculture

1.4.1 Current Status of Mechanization Development of Gobi Facility Agriculture

The mechanized development of Gobi facility agriculture is in its infancy. This process is full of opportunities and challenges. At present, the mechanization process shows an uneven development trend in different links. Significant progress has been made in some areas, while breakthroughs are still needed in other areas. Let's explore this development and its implications in detail.

In terms of matrix production equipment, production needs have been basically met. These equipment mainly include crushers, compost turners, loaders and bagging equipment. They play a key role in agricultural waste treatment and reuse. For example, a crusher is used to break waste such as crop straw into preparation for subsequent processing. In the fermentation process of agricultural waste products such as cow dung, straw, and bacterial residues, the pile turner is responsible for multiple tasks such as turning, dumping, overturning, moving and rolling to ensure the uniformity and fullness of the fermentation process. Loaders and bagging equipment further increase the efficiency of matrix handling and packaging. The application of these equipment not only improves the utilization rate of agricultural waste, but also provides high-quality planting substrate for Gobi facility agriculture, laying a good foundation for crop growth. The mechanization application in greenhouse management has developed rapidly and achieved remarkable results. The greenhouse rolling curtain and rolling film ventilation systems have been fully mechanized, with the mechanization rate reaching 100%. This means that the environment inside the greenhouse becomes more precise and efficient, which is beneficial to crop growth management. In addition, the application of integrated water and fertilizer equipment makes the irrigation and fertilization process more scientific and economical. The use of greenhouse snow plows greatly reduces the pressure caused by winter snow on the greenhouse and ensures the safety of the facility. It is worth noting that some new equipment, such as heating equipment for extreme low temperatures and lightweight material transport vehicles, have entered the trial and promotion stage. The application of these innovative equipment will further improve the risk resistance and production efficiency of Gobi facility agriculture. In some key aspects of greenhouse crops, mechanization development is still relatively lagging behind. For example, there is currently a lack of special equipment suitable for mechanized operations in areas such as seedling transplanting, crop harvesting, and straw and vegetable processing. The mechanization development of these links is slow. The main reasons may include: high technical difficulty, high investment in equipment research and development, and poor applicability. This unbalanced development situation results in these links still relying on a large amount of labor, which not only increases production costs, but also limits the improvement of overall production efficiency.

Faced with this uneven development of mechanization, we need to take a series of measures to promote comprehensive development:

1. Increase investment in research and development: Increase research and development funds for key links such as seedling transplanting and harvesting, and encourage cooperation between agricultural machinery companies and scientific research institutions to develop special equipment suitable for the characteristics of Gobi facility agriculture.
2. Promote pilot demonstrations: For new equipment that has been developed, such as extreme low temperature heating equipment, increase pilot demonstration efforts, collect practical application data, and continuously optimize and improve.
3. Cultivate professional talents: Strengthen the training of agricultural mechanization talents, improve the skill level of operators, and provide talent support for the development of mechanization.
4. Improve policy support: Develop targeted subsidy policies to encourage farmers to purchase and use new agricultural machinery and equipment, and lower the cost threshold for mechanized production.
5. Promote the integration of industry, academia and research: Encourage universities, research institutions and agricultural enterprises to cooperate, combine theoretical research with practical needs, and accelerate the transformation of technological achievements.
6. Pay attention to environmental protection and sustainability: In the process of mechanization development, we must pay attention to environmental protection, develop energy-saving and environmentally friendly equipment, and achieve sustainable development.

Although the development of agricultural mechanization in Gobi facilities is still in its infancy, significant progress has been made in some areas. In the future, we need to continue our efforts to comprehensively promote the mechanization process to improve production efficiency, reduce costs, increase output, and ultimately achieve the sustainable development of Gobi facility agriculture. This is not only related to the development of local agriculture, but also of great significance to my country's food security and ecological environment protection.

1.4.2 Application of Mechanization in Gobi Facility Agriculture

The demand for "vegetable basket" products in my country has increased, while water and soil resources are becoming increasingly scarce. The National Modern Facility Agriculture Construction Plan (2023–2030) proposes that, under the premise of protecting the ecology and not increasing the total amount of water use, various non-arable land resources should be reasonably utilized, and Gobi, deserts, etc. should be scientifically utilized to develop facility agriculture.

At the 2024 China Desert Facility Agriculture Science and Technology Innovation Academic Exchange and the Launch Meeting of Major Scientific and Technological Tasks of the Chinese Academy of Agricultural Sciences, Yu Jingquan, an academician of the Chinese Academy of Engineering, said in a video speech that the development of Northwest facility agriculture has advantages including rich non-arable land resources, low large-scale development costs, rich light and heat resources, good vegetable quality, low humidity, and few diseases, which will help solve the problem of arable land shortage and increase the income of farmers in the west.

Exploring the scientific use of Gobi, deserts, etc. to develop facility agriculture is one of the important measures to develop modern facility agriculture. This year's Central Document No. 1 proposed to promote the modernization and upgrading of facility agriculture.

The mechanization and automation of Gobi facility agriculture in my country is constantly improving. According to local environmental characteristics and agricultural needs, various regions have developed and applied a variety of agricultural machinery and equipment with strong adaptability and high reliability. Gansu Province, as one of the important areas of Gobi facility agriculture in my country, has made remarkable progress in the application of mechanical equipment. In terms of Gobi greenhouse rolling curtain machinery, the XJL-100-1 electric rolling curtain machine produced by Jiuquan Zhulong Machinery Manufacturing Co., Ltd. and the XJL-120 electric rolling curtain machine produced by Jiuquan Linde Machinery Manufacturing Co., Ltd. have been widely used. These two rolling curtain machines have shown good adaptability and reliability in actual use and have been well received by users. In terms of greenhouse leek harvesting, the 4GLZ-1 leek harvester produced by Jiuquan Linde Machinery Manufacturing Co., Ltd. has been put into use in Suzhou District and Yumen City, and also shows excellent adaptability and reliability. In terms of the application of water-fertilizer integrated technology, the 8GW-90 (60, 30)/4 agricultural irrigation water and fertilizer filtration head system developed by Gansu Dayu Water Saving Co., Ltd. has passed the agricultural machinery promotion and appraisal and will be put into use soon. The application of this system will greatly improve the utilization efficiency of water resources and fertilizers, and provide important support for the sustainable development of Gobi facility agriculture. In response to the problem of low temperature and cold damage reduction in Gobi greenhouses, the 5LS-20 biomass pellet hot air furnace developed by Jiuquan Zhongtian New Energy Technology Co., Ltd. was tested in the Dongdongtan Gobi Agricultural Demonstration Park in Suzhou District, with good results. The application of this hot air furnace will effectively solve the adverse effects of low temperatures in winter in the Gobi region on crop growth, extend the crop growth cycle and increase yields. In terms of light and simplified material transportation in greenhouses, five light and simplified material transport vehicles developed by Jiuquan Zhulong Machinery Manufacturing Co., Ltd. were installed in the Dongdongtan Gobi Agricultural Demonstration Park in Suzhou District. The use of this transport vehicle will greatly improve the efficiency of material transportation in greenhouses and reduce labor intensity. In Jiuquan City, Gansu Province, in addition to the equipment mentioned above, some new agricultural machinery is in the trial stage. The Gobi greenhouse intelligent supplementary lighting system developed by Jiuquan Agricultural Science Research Institute in cooperation with local enterprises is being tested in the Gobi Agricultural Demonstration Park in Suzhou District. The system can automatically adjust the switch and light intensity of the supplementary light according to the photosynthesis needs of crops and the natural light intensity, which is expected to increase crop yield and quality.

Greenhouse roller shutter machine

Intelligent fill light system

As another important Gobi facility agriculture region, Xinjiang has also made significant progress in mechanization and automation. The intelligent greenhouse environmental control system developed by Xinjiang Agricultural University in cooperation with local enterprises has been widely used in Gobi greenhouses in Turpan, Hami and other places. The system can automatically adjust the temperature, humidity, lighting and other parameters in the greenhouse according to crop growth needs and environmental conditions, greatly improving the accuracy and efficiency of greenhouse management. In the Shihezi region of Xinjiang, the Gobi Desert drip irrigation system developed by the School of Agriculture of Shihezi University has been promoted and applied. This drip irrigation system is optimized for the soil characteristics and crop water requirements in the Gobi region. It not only saves water significantly, but also effectively prevents soil salinization. Currently, the system has been installed in more than 1,000 acres in the Gobi agricultural demonstration zone around Shihezi.

Gobi facility agriculture in the Inner Mongolia Autonomous Region is also developing rapidly. In Alxa League, the Gobi greenhouse intelligent ventilation system developed by Inner Mongolia Agricultural University in cooperation with local enterprises has been applied in multiple facility agricultural parks. The system can automatically adjust the opening and closing of the vents according to the temperature difference and wind speed between the inside and outside of the greenhouse, effectively improving the internal environment of the greenhouse and improving the quality of crop growth. The Gobi Agricultural Demonstration Park in Baotou City, Inner Mongolia, has introduced the Gobi Greenhouse Intelligent Irrigation and

Fertilization Integrated Machine developed by the Beijing Agricultural Intelligent Equipment Technology Research Center. This equipment can automatically perform precise irrigation and fertilization based on soil moisture and crop growth conditions, greatly improving water and fertilizer utilization efficiency and reducing the need for manual operations.

Qinghai Province also has unique development in the mechanization of agricultural facilities in the Gobi Desert. The plateau Gobi greenhouse insulation system developed by the Qinghai Provincial Agricultural and Animal Husbandry Machinery Research Institute has been applied in the Gobi Agricultural Demonstration Zone of Haixi Mongolian and Tibetan Autonomous Prefecture. The system uses multi-layer insulation materials and intelligent control technology, which can effectively cope with the characteristics of large temperature differences between day and night and strong ultraviolet rays in plateau areas, and provide a stable growth environment for crops.

The Gobi facility agriculture in Ningxia Hui Autonomous Region is also actively introducing and applying new agricultural machinery. In the Gobi Agriculture Demonstration Park in Shapotou District, Zhongwei City, Ningxia, the Gobi greenhouse intelligent dehumidification system developed by the Ningxia Academy of Agricultural and Forestry Sciences is being tested and applied. This system can effectively control the humidity in the greenhouse, reduce the occurrence of pests and diseases, and improve the crop growth environment.

In the edge areas of the Mu Us Desert in Yulin City, Shaanxi Province, Gobi facility agriculture has developed rapidly. The special greenhouse covering system for desertified areas developed by the Shaanxi Agricultural Machinery Research Institute has been applied locally. The system adopts special materials and structural design, which can effectively resist the invasion of sandstorms and protect the internal environment of the greenhouse.

Gobi facility agriculture is being developed in the coal mining subsidence area of Datong City, Shanxi Province, as an important approach to ecological restoration and industrial transformation. The Gobi Greenhouse Intelligent Monitoring System for Mining Areas developed by Shanxi Agricultural University in cooperation with local enterprises has been put into use in the Gobi Agriculture Demonstration Park in Yungang District, Datong City. The system can monitor soil settlement and greenhouse structural stability in real time to ensure the safe production of facility agriculture.

Intelligent water and fertilizer irrigation machine

Greenhouse insulation system

The level of mechanization and automation of Gobi facility agriculture is constantly improving. Various regions have developed and applied a variety of agricultural machinery and equipment with strong adaptability and high reliability based on local environmental characteristics and agricultural needs. The application of these equipment not only improves production efficiency and reduces labor intensity, but also significantly improves the crop growth environment and increases yield and quality. However, many new types of equipment are still in the experimental stage, and their long-term adaptability and reliability need to be further verified in actual production. In the future, with the further application of technologies such as the Internet of Things, big data, and artificial intelligence, the mechanization and intelligence level of Gobi facility agriculture is expected to be further improved. For example, drones and satellite remote sensing technology are used to monitor large-area Gobi agriculture, machine learning algorithms are used to optimize irrigation and fertilization strategies, and robotic technology is used to realize automated management and harvesting in greenhouses. The application of these technologies will provide strong support for the sustainable development of Gobi facility agriculture and promote the development of Gobi agriculture in a more efficient, precise and environmentally friendly direction.

1.5 Challenges of Automation Operations in Desert Gobi Facilities

There are many challenges in the automation of desert Gobi facilities, which stem from the unique geographical environment, climatic conditions and resource constraints in the region. Understanding and responding to these challenges is essential to promoting the sustainable development of desert Gobi facility agriculture. As the foundation of desert Gobi facility agriculture, the design and management of greenhouses directly affect the efficiency and sustainability of the entire production system. In this special environment, the yellow sand matrix cultivation mode, as an innovative planting method, is becoming a hot topic of research. This cultivation mode makes full use of the abundant local sand resources, and also requires supporting mechanization and automation technology support.

The design and test of the yellow sand matrix self-propelled trenching and film covering machine in the desert Gobi greenhouse is a key link in promoting this cultivation mode. Although in the greenhouse environment, the machine does not have to face the extreme weather of the desert directly, factors such as high temperature, high humidity and limited space still need to be considered. The compactness and flexibility of the machine need to be considered during the design process to ensure that it can operate freely in the narrow greenhouse passages. At the same time, attention should be paid to controlling noise and emissions to avoid affecting the environment in the greenhouse. The test phase needs to be carried out in an actual greenhouse environment to evaluate the operation quality, efficiency and impact of the machine on the greenhouse environment, and continuously optimize the design based on the test results.

The greenhouse environment provides relatively stable lighting conditions for the application of intelligent identification and transplanting technology of seedlings in greenhouses, which is conducive to the accuracy of image recognition. However, the high humidity environment in the greenhouse may affect electronic equipment, and moisture-proof measures need to be considered in the design. Transplanting technology needs to adapt to the narrow space and the characteristics of the yellow sand substrate in the greenhouse, while also considering the interference of surrounding plants. This study needs to combine technologies such as image processing, artificial intelligence and precision mechanical control to develop an intelligent transplanting system suitable for greenhouse environments. In order to transform the above research results into practical applications, the design and test of the intelligent transplanting device for seedlings in greenhouses has become the next important research direction. This device needs to be able to move flexibly in the narrow passages of the greenhouse and accurately complete the grasping, transportation and planting of seedlings. The high temperature and high humidity environment in the greenhouse needs to be considered during the design process to ensure the long-term reliability of the device. At the same time, the interference of other equipment and plants in the greenhouse also needs to be considered. The test phase needs to be carried out in an actual greenhouse environment to evaluate the operating efficiency of the device, the survival rate of seedlings and the impact on the greenhouse environment.

In greenhouse environments, water and fertilizer management has become more sophisticated and complex. The study of the water and fertilizer demand model for fruits and vegetables in desert Gobi greenhouse facilities needs to consider the microclimate changes in the greenhouse, the characteristics of the yellow sand matrix, and the growth characteristics of different crops. This study requires the deployment of a precise sensor network in the greenhouse to monitor environmental parameters such as temperature, humidity, light, CO_2 concentration, as well as soil moisture and nutrient content in real time. By establishing an accurate water and fertilizer demand model, accurate management of each area and each crop in the greenhouse can be achieved. Based on the above water and fertilizer demand model, the design and optimization of the greenhouse intelligent water and fertilizer integrated system has become a key task. This system needs to be able to automatically adjust the water and fertilizer supply according to the environmental conditions and crop growth status of different areas in the greenhouse. The system design needs to consider the space limitations in the greenhouse and develop compact equipment. At the same time, it is also necessary to consider how to work with other greenhouse systems (such as ventilation, shading, etc.) to achieve optimal control of the overall environment. Pest and disease control faces new challenges. High temperature and high humidity environments are conducive to the reproduction of certain pests and diseases, while closed spaces limit the introduction of natural enemies. Therefore, the study of spray agent models in greenhouses is particularly important. This study needs to consider the air flow characteristics in the greenhouse, the spatial distribution of crops, and the propagation characteristics of different pests and diseases, and establish a spray agent model suitable for the greenhouse environment. At the same time, it is also necessary to study the negative impact of the agent on the greenhouse environment and crops. The design of the greenhouse intelligent sprayer enables flexible movement in the narrow space of the greenhouse, and automatically adjusts the spraying amount and spraying method according to the pest and disease conditions in different areas. The high humidity environment in the greenhouse needs to be considered during the design process to ensure the waterproof performance of the equipment. At the same time, it is also necessary to develop an accurate navigation system so that the sprayer can autonomously avoid other equipment and plants in the greenhouse.

The mechanized operation mode of greenhouse facility agriculture integrates the above technologies organically to form a complete production system. Develop an intuitive user interface and control system so that operators can easily monitor and adjust various equipment. At the same time, it is also necessary to develop an intelligent decision-making support system, as well as a remote monitoring and operating system so that experts can provide technical support to the site at any time and help managers formulate the best production strategy. This study designs a mechanized operation system suitable for the greenhouse yellow sand matrix cultivation mode from the entire process of planting preparation to harvesting. Fully consider the spatial limitations and environmental characteristics of the greenhouse, realize the coordinated operation of various equipment, and form a set of efficient, precise and sustainable greenhouse facility agricultural production models. The use of data collection and analysis has become more important and complex in the greenhouse environment. It is necessary to deploy a full range of sensor networks in the greenhouse to monitor environmental parameters, crop growth conditions, equipment operation status, etc. in real time. These data are not only used to guide daily production management, but also to establish accurate growth models and prediction systems. Researchers need to develop high-reliability sensors that can adapt to the high temperature and high humidity environment of the greenhouse, as well as a powerful data processing and analysis platform. Through in-depth and comprehensive research, a mechanized and automated technology system suitable for the yellow sand matrix cultivation mode of greenhouses in desert Gobi areas has been established.

Make full use of the advantages of the greenhouse environment, overcome its limitations, and achieve efficient, precise and sustainable agricultural production. It can not only greatly improve the agricultural production efficiency and product quality in desert Gobi areas, but also save water resources, reduce environmental pollution, and make important contributions to local economic development and ecological environment improvement. The application of the results will provide strong technical support for the development of greenhouse facility agriculture in desert Gobi areas, and promote agricultural production in these areas to develop in an efficient, precise and sustainable direction. At the same time, these technologies and experiences can also be extended to greenhouse agriculture in other similar arid and semi-arid areas, providing new solutions to the food security problems facing the world. Through continuous innovation and practice, greenhouse facility agriculture in desert Gobi is expected to become one of the important directions for future agricultural development, providing new ideas and methods for mankind to cope with global challenges such as climate change and resource shortages.

1.6 Reasons for Developing Automated Operations in Desert Gobi Facilities

The automation of facility agriculture in desert Gobi is not only an inevitable trend of technological progress, but also an inevitable choice to cope with severe environmental challenges and improve agricultural production efficiency. In this extremely arid and harsh land, traditional agricultural production methods face huge challenges. First, resources are scarce in desert Gobi areas, especially water resources are extremely scarce, which requires agricultural production to accurately control the use of every drop of water to achieve the highest water resource utilization efficiency. The automated irrigation system can accurately control the irrigation time and water volume according to multiple parameters such as soil moisture, crop growth stage and weather conditions, greatly improve the utilization rate of water resources and reduce unnecessary waste. Secondly, the temperature difference between day and night in desert Gobi areas is large, and the climate changes dramatically, which puts extremely high demands on the control of crop growth environment. The automated environmental control system can monitor environmental factors such as temperature, humidity, and light in real time, and automatically adjust the internal environment of the greenhouse according to the growth needs of crops to provide the best growth conditions for crops. In addition, labor is scarce and costly in desert Gobi areas, and automated operations can greatly reduce manpower requirements and reduce production costs. For example, automatic seeders, transplanters and harvesters can replace a large number of manual operations, which not only improves operation efficiency, but also ensures the consistency of operation quality. Automation can also reduce human operational errors and improve production accuracy and stability. In terms of fertilizer and pesticide use, the automated distribution system can accurately control the amount and time of fertilizer application based on soil nutrient conditions and crop growth needs, ensuring both crop nutrient supply and avoiding environmental pollution caused by excessive fertilization. Similarly, the automated pest and disease monitoring and control system can promptly detect and accurately handle pest and disease problems, minimize the use of pesticides, and protect the ecological environment.

In extreme environments such as deserts and Gobi, automated operations are not only a means to improve production efficiency, but also the key to ensuring the sustainability of agricultural production. The continuous operation of the automated system can collect and analyze a large amount of data, which is of great value in optimizing production strategies, predicting yields, and managing risks. Through big data analysis and artificial intelligence technology, farm managers can make more scientific and accurate decisions to further improve production efficiency and economic benefits. For example, by analyzing historical meteorological data and crop growth data, planting plans can be optimized to select crop varieties and planting times that best suit local conditions. Automated operations can also improve the quality and safety of agricultural products. By precisely controlling the growth environment and the use of inputs, healthier and safer agricultural products can be produced to meet consumers' growing demand for high-quality food. In terms of pest control, automated monitoring systems can detect problems early and achieve precise application of pesticides, which not only reduces the use of pesticides, but also reduces the risk of pesticide residues in agricultural products. In addition, automated operations can also improve resource utilization efficiency and reduce environmental pollution. For example, through circulating hydroponic technology and precision fertilization systems, water resource consumption and fertilizer loss can be greatly reduced, reducing the impact on the surrounding environment. In terms of energy use, the automation system can optimize energy distribution, such as automatically adjusting the electricity consumption time according to electricity price fluctuations and electricity demand, which not only saves energy costs but also reduces carbon emissions. In the long run, the promotion of automated operations will promote the development of the entire desert Gobi facility agriculture towards intelligence and precision, laying the foundation for building a sustainable modern agricultural system. It can not only increase the yield per unit area, but also optimize resource allocation, reduce environmental pressure, and contribute to the ecological restoration and environmental protection of the desert Gobi region. At the same time, the promotion of automated operations will also drive the development of related industries, such as agricultural machinery manufacturing, software development, sensor technology, etc., injecting new vitality into local economic development. In addition, automated operations can also increase the attractiveness of agricultural production, attract more young people to participate in modern agriculture, and solve the problem of aging rural labor. Through remote control and monitoring systems, farm managers can realize centralized management of multiple production bases, greatly improving management efficiency and decision-making level. This intensive and large-scale management model can not only reduce production costs, but also improve the market competitiveness of agricultural products. Finally, the promotion of automated operations will drive the development of desert Gobi facility agriculture towards high-tech and high value-added directions, improve the overall competitiveness and risk resistance of agriculture, and make important contributions to ensuring national food security and promoting regional economic development.

1.7 Future Development Trends and Prospects

The future development trend and outlook of greenhouse facility agriculture in desert Gobi is a field full of opportunities and challenges. With the continuous advancement of technology and the in-depth understanding of the concept of sustainable development, this field will show a diversified, intelligent and ecological development trend.

Intelligence is one of the core development directions of greenhouse facility agriculture in desert Gobi in the future. The widespread application of artificial intelligence (AI) and Internet of Things (IoT) technologies will completely change the traditional agricultural production model. The intelligent sensor network will cover the entire greenhouse and monitor various parameters such as temperature, humidity, light, CO_2 concentration, soil moisture and nutrient content in real time. These data will be input into the intelligent control system, analyzed and decided by machine learning algorithms, and the greenhouse environment will be automatically adjusted to optimize water and fertilizer supply. In the future, greenhouses will be fully automated, and the entire process from sowing, transplanting to harvesting will be completed by intelligent robots, greatly reducing manpower requirements and improving production efficiency. The application of precision agriculture technology in greenhouse environments will be more in-depth. Crop growth models based on big data analysis and artificial intelligence will become more accurate, able to predict crop growth conditions and yields, and guide precise management decisions. The pest and disease early warning system will combine meteorological data and crop growth data to predict possible pests and diseases in advance and implement a prevention-oriented plant protection strategy. The precision fertilization system will supply the required nutrients quantitatively and regularly according to the real-time needs and growth stages of crops, maximize fertilizer utilization efficiency, and reduce environmental pollution. The application of blockchain technology will completely change the supply chain management of agricultural products. The entire process from planting to sales will be recorded on the blockchain to ensure product traceability and transparency. This is not only conducive to food safety management, but also helps farmers get fairer prices.

Energy management will become an important research direction for greenhouse facility agriculture in the future. The use of renewable energy such as solar energy and wind energy will be more extensive and efficient. New solar photovoltaic glass will be directly used in greenhouse construction, which can not only provide the required energy for greenhouses, but also adjust the light intensity and optimize the crop growth environment. The utilization technology of geothermal energy and shallow geothermal energy will continue to mature, providing a stable heat source for greenhouses. In addition, the application of smart microgrid technology will achieve efficient distribution and utilization of energy, ensuring stable operation of greenhouses around the clock. Gene editing and molecular breeding technologies will cultivate crop varieties that are more adapted to the desert Gobi greenhouse environment. These new varieties will have stronger stress resistance, higher resource utilization efficiency and better quality. At the same time, the development of functional agricultural products will become a research hotspot, such as vegetables and fruits rich in specific nutrients or with health care functions. Artificial light source technology will achieve major breakthroughs. The new generation of LED lights is not only more energy-efficient, but also can precisely control the spectrum to optimize crop photosynthesis. Programmable dynamic lighting systems will simulate the changing laws of natural light to create the best growth environment for crops. In addition, biochar technology will play an important role in improving sandy soils. Using plant waste in greenhouses to make biochar can not only improve soil fertility and water retention capacity, but also fix a large amount of carbon, which helps to mitigate climate change.

Vertical farming technology will be fully applied in greenhouses to maximize the use of limited space. Multi-layer three-dimensional planting systems will greatly increase the yield per unit area. Combined with LED supplementary lighting technology, multiple microclimate zones can be created inside the greenhouse to meet the growth needs of different crops.

The development of remote control and automation technology will make "unmanned greenhouses" possible. Through 5G networks and edge computing technology, remote monitoring and management of greenhouses can be achieved. This can not only reduce labor costs, but also apply expert wisdom to agricultural production in remote areas.

These development trends indicate that desert Gobi greenhouse facility agriculture will usher in a new era of intelligence, precision, ecology and diversification. Through technological innovation and system integration, desert Gobi areas are expected to become efficient and sustainable modern agricultural production bases, making important contributions to solving global food security issues and addressing climate change challenges. However, to realize these beautiful visions, continuous scientific research investment, policy support and interdisciplinary cooperation are still needed. The future desert Gobi greenhouse facility agriculture will be a comprehensive high-tech industry integrating agricultural science, information technology, materials science, environmental science and other disciplines, providing valuable experience for human beings to explore sustainable survival in extreme environments.

Chapter 2
Study on the Cultivation Model of Yellow Sand Substrate in Desert Gobi

Zhanming Tan, Yalong Song, Jianfei Xing, Long Wang, Jiahui Xu, Xufeng Wang, and Xiaohang Liu

2.1 Introduction

In the contemporary context of global climate change and increasingly scarce resources, sustainable agricultural development has become one of the key research priorities worldwide. Deserts and Gobi regions cover extensive areas of the Earth's surface. Due to their extreme climates and barren soil conditions, these areas have long been considered unsuitable for traditional agriculture. However, with the rising global food demand and the reduction of arable land resources, the development and utilization of these marginal lands, especially by leveraging their unique yellow sand substrates for agricultural cultivation, have become particularly important.

Z. Tan · Y. Song · J. Xing (✉) · L. Wang · J. Xu · X. Wang
Modern Agricultural Engineering Key Laboratory at Universities of Education Department of Xinjiang Uygur Autonomous Region, Alar 843300, China
e-mail: 591101353@qq.com

Xinjiang Production and Construction Corps (XPCC) Key Laboratory of Utilization and Equipment of Special Agricultural and Forestry Products in Southern Xinjiang, Alar 843300, China

L. Wang
e-mail: 120140002@taru.edu.cn

Z. Tan
College of Horticulture and Forestry, Tarim University, Alar 843300, China

Y. Song · J. Xing · L. Wang · J. Xu · X. Wang
College of Mechanical and Electrical Engineering, Tarim University, Alar 843300, China

X. Liu
Key Laboratory of Smart Agriculture System Integration, Ministry of Education, Beijing 100083, China

Key Laboratory of Agricultural Information Acquisition Technology, Ministry of Agriculture and Rural Affairs, Beijing 100083, China

C. Hu et al. (eds.), *New Technologies Applied in Gobi Desert Facility Agriculture*, Smart Agriculture 11, https://doi.org/10.1007/978-981-96-1031-0_2

Yellow sand substrates possess high air permeability and good drainage capability, which offer potential advantages in certain agricultural applications. Additionally, these physical properties of yellow sand can reduce the incidence of certain plant diseases, providing a unique growth environment for specific crops. However, yellow sand also has poor nutrient retention, necessitating the development of new cultivation techniques and management strategies to support crop growth effectively.

The aim of this chapter is to explore and develop agricultural cultivation models on yellow sand substrates in desert and Gobi regions, with a particular focus on the comparative study of north–south versus east–west cultivation models. This research not only aims to provide technical support for desert greening and ecological restoration but also seeks to contribute to global food security. By conducting a detailed analysis of these two cultivation models, this study hopes to identify the most suitable agricultural practices under desert and Gobi conditions, thereby enhancing crop production efficiency and quality.

Investigating the specific effects of north–south and east–west cultivation models will aid in understanding how environmental factors such as sunlight, temperature, and humidity influence crop growth under extreme climatic conditions. For instance, north–south orientation may favor certain crops by optimizing sunlight exposure, whereas east–west orientation might be better for managing wind and moisture. Through experiments and simulations, this study will evaluate the performance of these cultivation models in practical applications, providing scientific recommendations for future desert agriculture.

Furthermore, the study will focus on developing techniques to improve yellow sand substrates, such as the incorporation of organic matter and nutrients, to enhance their nutrient retention and crop growth potential. By integrating knowledge from soil science, plant physiology, and environmental engineering, this study aims to create a sustainable agricultural system for desert and Gobi regions.

In summary, the objectives and significance of this chapter are to: explore and optimize the feasibility of using yellow sand substrates for agricultural cultivation in desert and Gobi regions; compare and evaluate the performance and adaptability of north–south versus east–west cultivation models under different environmental conditions; develop crop cultivation techniques and soil improvement methods suitable for extreme environments; and provide theoretical and practical support for the agricultural development of marginalized lands globally, thereby promoting ecological restoration and sustainable food production.

2.2 Characteristics of Yellow Sand Substrates

2.2.1 *Physical Properties of Yellow Sand*

Yellow sand, primarily formed from Aeolian deposits, has physical properties that are crucial for determining its application in agricultural production. The texture of yellow sand is relatively coarse, mainly composed of sand particles with diameters ranging from 0.05 to 2 mm. This texture endows yellow sand with high porosity and good water permeability, allowing for the rapid drainage of excess water, thereby reducing moisture retention around the roots and lowering the incidence of root diseases. However, this high permeability also leads to poor water retention, which can limit plant growth, especially under drought conditions, due to the rapid loss of moisture.

2.2.2 *Chemical Properties of Yellow Sand*

Chemically, yellow sand substrates are typically nutrient-poor, with especially low organic matter content, due to the lack of organic accumulation during the weathering process. The pH of yellow sand is generally neutral to slightly alkaline, which can be advantageous for certain alkaline-loving plant species. However, for most crops, it may be necessary to adjust the pH and nutrient levels by adding organic materials or fertilizers to meet the growth requirements of the plants. Additionally, the content of essential trace elements such as potassium, calcium, and magnesium is relatively low in yellow sand. Since these elements are critical for plant growth, their supplementation is essential when using yellow sand as a cultivation substrate.

2.2.3 *Impact of Yellow Sand on Plant Growth*

The physical and chemical properties of yellow sand directly affect plant growth. Its excellent aeration and drainage properties provide a favorable growth environment for plant roots, particularly for root systems that thrive in well-aerated conditions, such as some xerophytic shrubs and succulents, which can exhibit good growth performance in this substrate. However, the poor water and nutrient retention capabilities of yellow sand mean that, without proper management, plants may face severe water and nutrient deficiencies. Therefore, when cultivating with yellow sand, additional irrigation and fertilization measures are usually required to ensure normal plant growth. For example, the application of drip and micro-irrigation techniques can effectively improve water use efficiency, while the use of organic fertilizers and slow-release chemical fertilizers can provide long-term, stable nutrient supply.

Understanding and leveraging these characteristics of yellow sand substrates can significantly enhance the likelihood of plant survival and productivity in extreme environments, such as desert and Gobi regions. By improving yellow sand substrates and integrating modern agricultural technologies, it is possible to develop efficient and sustainable cultivation models tailored for arid and semi-arid areas.

2.3 Overview of Cultivation Models

The cultivation model using Gobi yellow sand substrates represents an innovative agricultural technology that utilizes yellow sand from Gobi and desert regions as a substrate for agricultural production through specialized cultivation techniques. This method not only helps improve and utilize marginal lands but also effectively prevents desertification processes.

(1) **East–West Orientation Greenhouse Cultivation Model**

In the Gobi yellow sand substrate, the East–West orientation greenhouse cultivation model optimizes light reception by aligning the main axis of the greenhouse in an East–West direction. This orientation helps capture low-angle sunlight during the morning and evening, enhancing light efficiency, particularly in the spring and fall seasons. Additionally, this layout reduces the intensity of direct overhead sunlight during the hot summer months, thereby lowering the internal temperature of the greenhouse and mitigating the adverse effects of high temperatures on plant growth.

(2) **North–South Orientation Greenhouse Cultivation Model**

The North–South orientation greenhouse cultivation model aligns the main axis of the greenhouse in a North–South direction. This layout ensures that plants inside the greenhouse receive more uniform sunlight exposure throughout the day as the sun's trajectory changes. In the summer, this orientation helps reduce direct sunlight during midday, alleviating heat stress; in the winter, it maximizes sunlight exposure, increasing temperature and extending the growing season.

(3) **Substrate Bag Planting**

Substrate bag planting is a flexible cultivation method suitable for Gobi and desert regions with limited space or complex terrain. By filling specially designed substrate bags with yellow sand and other nutrients, crops can be planted directly. This method is easy to manage and relocate, allowing precise control over water and nutrient supply to the crops. The use of substrate bags reduces the risk of soil-borne diseases and allows for custom formulations of the substrate based on crop requirements.

(4) **Elevated Cultivation Systems**

Elevated cultivation systems or raised planting beds are particularly useful in Gobi and desert regions. These systems elevate plants above the ground, reducing the impact of ground heat radiation on crops while facilitating easier management and harvest. These systems typically incorporate drip irrigation technology to maximize water resource efficiency, preventing excessive irrigation and minimizing water evaporation and waste.

2.4 North–South Orientation Cultivation Model

Based on the practical issues encountered in protected agriculture and the environmental conditions in southern Xinjiang that are suitable for this cultivation model, the study proposes a North–South orientation cultivation model using yellow sand substrates. This model effectively addresses the challenges faced in protected agriculture. The steps of this cultivation model are as follows:

S1: As shown in Fig. 2.1, dig planting trenches within the protected environment, each trench being 8 m long, 0.4 m wide, with an internal depth of 0.3 m and an external depth of 0.35 m. As depicted in Fig. 2.2, maintain an interval of 0.8 m between each planting trench. According to Fig. 2.3, leave a drainage ditch with a depth of 0.35 m on the outer side of the planting trench. Cover the drainage ditch with plastic mulch and secure it to facilitate the drainage of excess water and fertilizers during crop management. Additionally, establish an underground liquid storage pool at the end of the protected cultivation area to collect and recycle excess water and fertilizers.

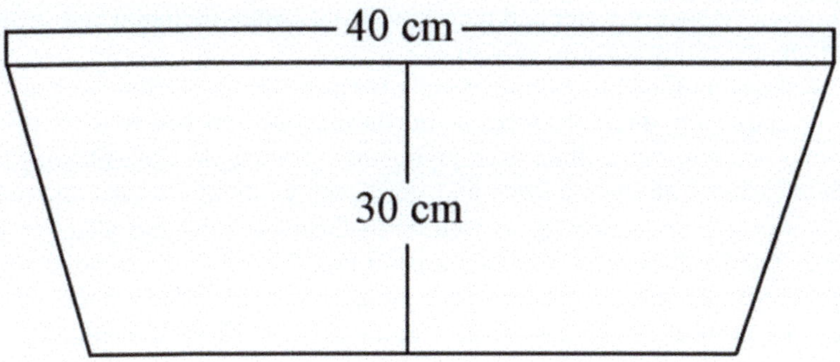

Fig. 2.1 Cross-section of planting trench

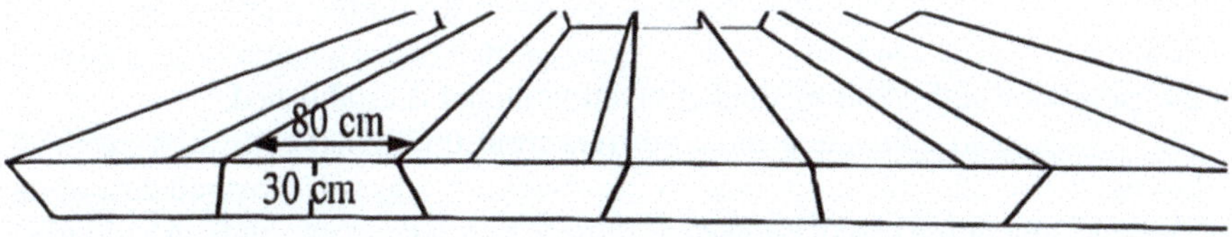

Fig. 2.2 Planting trench

Fig. 2.3 Liquid drainage channel

S2: As shown in Fig. 2.4, after digging the planting trenches as required, thoroughly irrigate the trenches and observe whether the water flows from the inner side to the outer side of the trenches. If the water does not flow as expected, adjust the soil within the trenches until a smooth flow from the inner side to the outer side is achieved. As illustrated in Fig. 2.5, uniformly cover the planting trenches with HDPE film, wrapping both the inner and outer sides of the trenches securely. Do not cut the high-density PE film between the two planting trench paths; instead, connect and cover them seamlessly, securing the film in place. The high-density PE film effectively isolates the soil beneath the planting trenches from alkali and salt intrusion, prevents the loss of water and nutrients, avoids water pollution and secondary salinization, and promotes crop growth.

S3: As shown in Fig. 2.6, thoroughly mix furnace slag and yellow sand at a mass ratio of 3:5 and evenly spread the mixture into the planting trench. Ensure the substrate is not too loose during spreading. After laying the substrate, flood the planting trench with water mixed with disinfectants and sterilizers, washing the substrate twice to remove pathogens and diseases, as well as to flush out salts and alkali from the furnace slag and yellow sand. As illustrated in Fig. 2.7, lay drip irrigation tapes on top of the substrate and cover them with a plastic film. This setup reduces substrate moisture evaporation, conserves moisture and nutrients, lowers the humidity inside the greenhouse, decreases the occurrence of pests and diseases, and extends the lifespan of the drip irrigation tapes. During the use of the subsurface drip irrigation system, apply water frequently in small amounts, and administer fertilizers in small, multiple doses. This is important because excess water in the planting trenches, if not drained properly, can lead to waterlogging, which is detrimental to crop growth. At this point, the entire yellow sand cultivation model is established.

The yellow sand cultivation model differs from traditional cultivation methods and trough substrate cultivation models in several key ways. The overall cost of the yellow sand cultivation model is lower, as furnace slag and yellow sand can be sourced locally, and the process is simple to operate, making it accessible for the general

Fig. 2.4 Monitoring water flow direction in the planting trench

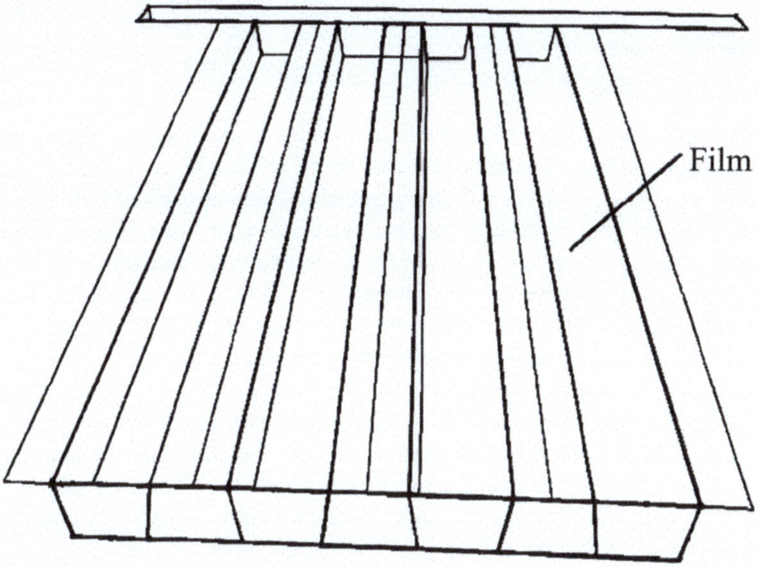

Fig. 2.5 Laying HDPE film in the planting trench

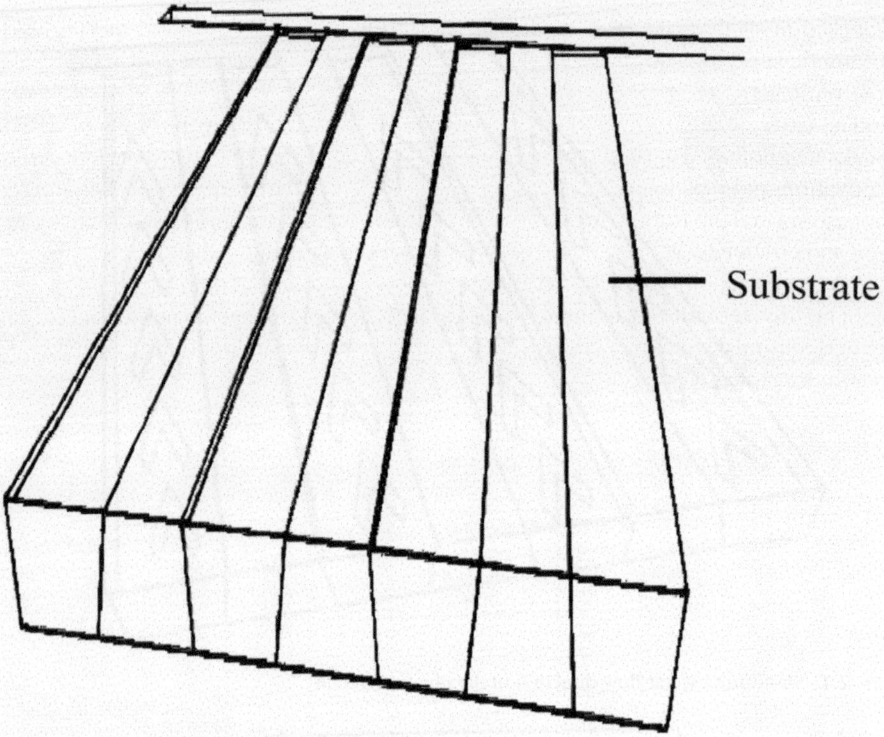

Fig. 2.6 Laying substrate in the planting trench

public. This planting model offers greater convenience and speed, resulting in time, labor, and effort savings. The substrate composed of yellow sand and furnace slag helps stabilize root temperature, thereby reducing fluctuations and aiding in water retention. Furthermore, the yellow sand and furnace slag substrate overcomes issues of soil monoculture obstacles, allowing for the continuous cultivation of the same genus and species of vegetables over multiple years. This facilitates ongoing management by technical personnel on specific vegetable crops, enabling the accumulation of extensive cultivation and management experience, which in turn promotes the rapid development of vegetable planting technology. The use of high-density polyethylene (HDPE) film isolates the substrate from the external environment, preventing issues such as soil salinization and alkalization, and protecting the surrounding soil and groundwater from contamination by nutrient solutions. The drip irrigation tapes, concealed beneath the plastic film, enhance their longevity and indirectly reduce production costs. Employing subsurface drip irrigation technology achieves water conservation goals, enabling plant root systems to fully absorb necessary nutrients and oxygen from different soil depths.

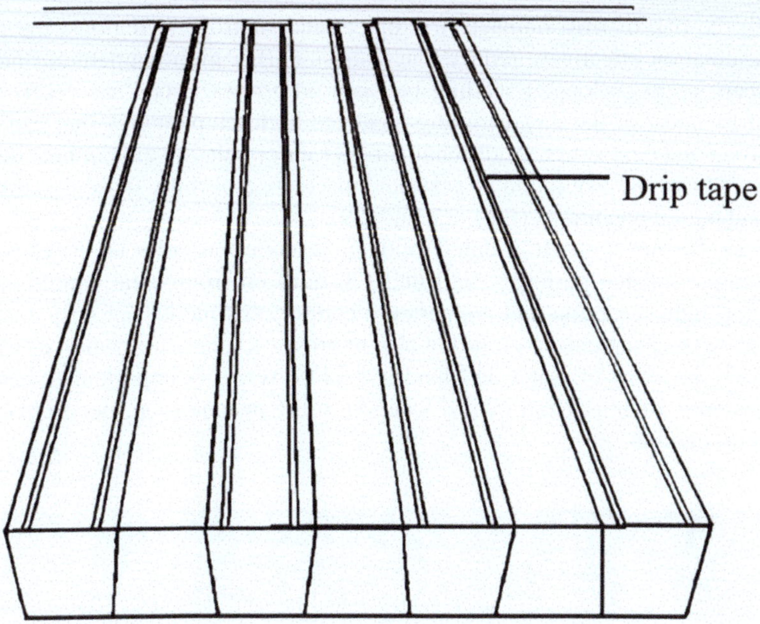

Fig. 2.7 Placing drip irrigation tape on top of the substrate

2.4.1 Environmental Adaptability Analysis: North–South Oriented Cultivation Model

The north–south oriented cultivation model exhibits significant advantages in light management. Due to the axial tilt of the Earth's rotation, crop rows oriented north–south receive uniform sunlight for the majority of the day. In summer, when the solar altitude is high, these crop rows can receive sunlight almost all day long without being shaded by adjacent rows, thus maintaining high photosynthetic efficiency. In winter or in high-latitude regions, the north–south orientation continues to demonstrate its benefits. With the sun at a lower angle, east–west oriented crop rows may suffer from severe shadow effects, resulting in some areas between the rows not receiving adequate sunlight. Conversely, north–south oriented rows can maximize the utilization of low-angle sunlight, especially during mornings and evenings, when direct sunlight penetrates the rows and reduces shaded areas, thereby enhancing overall photosynthetic efficiency.

The north–south oriented cultivation model also offers advantages in temperature management. In temperate and cold regions, moderate sunlight reception helps maintain suitable soil temperatures, promoting root development and microbial activity. The spacing between north–south oriented rows allows for air circulation, which helps mitigate the impact of extreme temperatures on crops. Particularly at night, improved air circulation can prevent rapid drops in ground temperature, thereby reducing the risk of frost damage. In tropical and subtropical regions, the north–south orientation effectively reduces heat stress during high-temperature periods. Since there is no prolonged shading between crop rows, crops can receive even sunlight for most of the day, avoiding localized high temperatures that can cause evaporation and heat stress. Additionally, better air circulation aids in heat dissipation and reduces the likelihood of pests and diseases, especially those that thrive in warm and humid conditions (Figs. 2.8 and 2.9).

Overall, the north–south oriented cultivation model provides an effective crop row arrangement that performs well under various environmental conditions. By optimizing light reception and temperature control, this model not only enhances the growing environment for crops but also improves resource utilization efficiency, particularly in terms of water and land use. However, it is important to consider specific factors such as geographical location, soil type, and crop species to ensure optimal production.

Fig. 2.8 North–South oriented cultivation for vegetables: Cucumber

Fig. 2.9 North–South oriented cultivation for vegetables: Tomato

2.5 East–West Oriented Cultivation Model

2.5.1 Concept and Application Background of East–West Orientation

Greenhouse cultivation technology, through artificial control of the cultivation environment, not only significantly increases crop yield and quality but also effectively extends the production season, reducing the impact of natural disasters on agricultural production. Among the key factors determining the energy efficiency and production benefits of a greenhouse is the choice of orientation. Traditionally, greenhouses have predominantly adopted a north–south orientation to ensure plants receive even sunlight at different times of the day. However, with in-depth research on different geographical locations, climatic conditions, and specific crop requirements, the east–west oriented greenhouse cultivation model has gradually demonstrated its unique advantages (Figs. 2.10 and 2.11).

The east–west oriented greenhouse cultivation model refers to a layout where the main axis of the greenhouse runs in an east–west direction. This layout can more effectively manage sunlight under certain latitudes and climatic conditions, particularly during the spring and autumn seasons when the sun is at a lower angle. In such cases, the east–west orientation allows plant leaves to receive more uniform sunlight, thereby enhancing photosynthetic efficiency and overall crop productivity. Additionally, during summer, the east–west orientation can reduce direct sunlight on crops, alleviating heat stress and helping to prevent physiological diseases caused

Fig. 2.10 East–West oriented cultivation model

Fig. 2.11 East–West oriented cultivation experiment: Aksu Naida Agriculture Co., Ltd.

by excessive heat. In winter, this orientation maximizes the capture of low-angle sunlight, increasing the temperature within the greenhouse and reducing heating costs.

Although the east–west orientation theoretically offers numerous advantages, its practical effectiveness is influenced by various factors, including the latitude of the geographical location, greenhouse structural design, choice of covering materials, and internal layout. Therefore, a comprehensive evaluation of the east–west oriented greenhouse cultivation model requires combining field trials and long-term monitoring data to thoroughly analyze its adaptability and production efficiency under different environmental conditions.

2.5.2 *Design Principles of East–West Oriented Greenhouses*

The East–West oriented greenhouse cultivation model optimizes light utilization efficiency and crop growth conditions through its unique structural design. This design principle is based not only on a deep understanding of the natural environment but also integrates the latest advancements in agricultural technology, particularly in light management and temperature control. The following section details the design principles of East–West oriented greenhouses, including their structural characteristics, light optimization, temperature regulation, and their impact on crop growth (Fig. 2.12).

Fig. 2.12 East–West oriented cultivation model—Liutuan, Aksu Prefecture, Xinjiang Uygur autonomous region

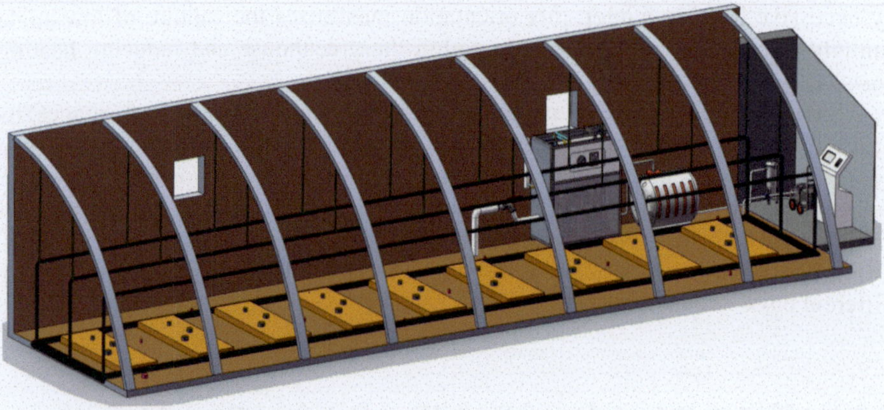

Fig. 2.13 Schematic diagram of greenhouse internal environment

In an East–West oriented greenhouse, the main axis is aligned in an east–west direction, allowing the north–south sides to receive uniform sunlight for most of the day. This structural design facilitates more even sunlight exposure on plant leaves, especially during the spring and autumn when the sun is at a lower angle, notably in the mornings and evenings. Compared to north–south oriented greenhouses, the east–west design reduces direct sunlight on crops at midday, helping to prevent photosynthetic saturation or photo inhibition that may occur under high-temperature and high-light conditions (Fig. 2.13).

Light is one of the fundamental conditions for plant growth. The East–West oriented greenhouse optimizes light distribution and utilization through specialized design. In such greenhouses, the roof and side walls are typically designed to accommodate the seasonal variations in solar altitude. By adjusting the light transmission properties and reflectivity of the covering materials, as well as incorporating internal reflective nets or white shading curtains, the amount and quality of light entering the greenhouse can be effectively managed. This light management strategy not only increases the available light energy for the plants but also balances the diurnal variations in light intensity, preventing the stress caused by excessively strong or weak light conditions.

Temperature is another critical factor affecting crop growth within greenhouses. In East–West oriented greenhouses, special consideration is given to ventilation and heat dissipation needs to address high-temperature issues during the summer. By configuring openable windows on the roof and sides, in combination with automatic or semi-automatic temperature control systems, the internal temperature and humidity of the greenhouse can be effectively regulated. Additionally, the use of insulating materials such as double-layer films is a common practice. These materials can reduce heat loss during winter and block some infrared radiation in the summer, maintaining the greenhouse temperature within an optimal range.

The East–West oriented greenhouse cultivation model provides a more ideal growth environment for crops by optimizing light and temperature management. This design ensures that crops receive adequate light throughout their growth cycle, while also mitigating the adverse effects of extreme climatic conditions. Consequently, compared to traditional North–South oriented greenhouses, the East–West oriented design more effectively promotes crop growth and development, enhancing both yield and quality, especially for high-value crops that are sensitive to light and temperature.

In summary, the East–West oriented greenhouse cultivation model optimizes key agricultural production factors through its unique structural design and management strategies, thereby improving the efficiency and sustainability of crop production. This design not only leverages the natural environment but also integrates modern technological applications, showcasing the trend towards high-tech, high-efficiency development in modern agriculture.

2.5.3 *Environmental Adaptability of East–West Oriented Greenhouses*

The East–West oriented greenhouse cultivation model is increasingly being adopted worldwide. A critical factor influencing its successful dissemination is its environmental adaptability. Whether this cultivation model can exhibit its inherent advantages under various geographical and climatic conditions is a subject of continuous scientific research and practical experimentation. This section delves into the adaptability of East–West oriented greenhouses under different climatic conditions, particularly in temperate, subtropical, and other unique environments, and examines their suitability for specific crops (Fig. 2.14).

In temperate regions, where there are distinct seasons and significant variations in solar altitude, East–West oriented greenhouses can maximize the utilization of the lower sun angles typical of spring and autumn. By optimizing the design of the sidewalls and roof structure, these greenhouses can enhance the uniformity and efficacy of light exposure. Additionally, the colder winters in temperate zones can be mitigated by increasing the solar reception area on the southern side of the greenhouse, which effectively raises internal temperatures and reduces heating costs. During summer, appropriate shading and ventilation designs can prevent the adverse impacts of high temperatures on crop growth.

In subtropical regions, characterized by abundant sunlight and higher temperatures, the East–West oriented greenhouse can effectively control excessive solar radiation through its structural design. This helps in avoiding rapid evaporation and heat stress on crops. In such regions, the East–West orientation optimizes light utilization, reduces the intensity of direct sunlight, and maintains a suitable internal climate through efficient ventilation systems, benefiting the growth of tropical and subtropical crops such as tomatoes and cucumbers.

Fig. 2.14 East–West oriented cultivation model—Tomato

In unique environments such as highland and coastal areas, East–West oriented greenhouses also demonstrate excellent adaptability. Highland regions, with their ample sunlight but large temperature fluctuations, benefit from the greenhouse's ability to maximize sunlight utilization while minimizing temperature variability. Coastal areas, which may face challenges such as strong winds, require greenhouses with enhanced structural stability and wind resistance. The East–West orientation in these areas ensures that the greenhouse can withstand such natural factors while maintaining an optimal growth environment.

Different crops have varying requirements for light and temperature. The East–West oriented greenhouse cultivation model can be adjusted in structure and internal configuration to meet these diverse needs. For leafy vegetables, which require less light than fruiting vegetables, the greenhouse can use shading nets to reduce high-intensity direct light and prevent growth inhibition caused by excessive light. Fruiting vegetables, which need more light to promote fruit development, benefit from the more uniform light distribution provided by East–West oriented greenhouses, encouraging overall growth and even fruit maturation. Ornamental and medicinal plants, which often have more specific environmental needs, can thrive in East–West oriented greenhouses equipped with precise environmental control technologies such as automatic temperature and humidity regulation systems, creating ideal growth conditions for these high-value crops (Fig. 2.15).

Fig. 2.15 East–West oriented cultivation model—Tomatoes

The environmental adaptability of East–West oriented greenhouses is a significant advantage in their design and application. Through flexible structural design and advanced environmental control technologies, this cultivation model can adapt to various climatic conditions and meet the growth needs of different crops. It showcases the potential of modern agricultural technology to enhance crop production efficiency and sustainability. As technology advances and practical experience accumulates, the design of East–West oriented greenhouses will become more refined, and their application in global agricultural production will be more widespread and profound.

2.5.4 Conclusion

The East–West oriented greenhouse cultivation model, as an innovative application of modern agricultural technology, primarily excels in optimizing light management and improving crop growth environments, although it also faces certain operational challenges. This section summarizes the core advantages and existing issues of East–West oriented greenhouses and offers recommendations for future research and practice.

High Light Utilization Efficiency: The design of East–West oriented greenhouses allows crops to receive relatively uniform light throughout the day. This is especially beneficial during the spring and autumn seasons when low-angle sunlight can more effectively reach the crops, enhancing photosynthetic efficiency.

Effective Temperature Regulation: This orientation helps reduce the intensity of direct overhead sunlight during hot summer months. Through appropriate shading and ventilation designs, the internal temperature of the greenhouse can be effectively controlled, mitigating the adverse effects of high temperatures and protecting crops from heat stress.

Energy Efficiency and Emission Reduction: By reducing reliance on heating and cooling equipment, East–West oriented greenhouses lower energy consumption, promoting green production. This is of significant importance for advancing sustainable agricultural practices.

Enhanced Yield and Quality: Balanced light exposure and suitable temperature and humidity conditions contribute to healthy crop growth, subsequently increasing both the yield and quality of agricultural products, thereby enhancing their market competitiveness.

However, several challenges also exist:

High Design and Construction Costs: Compared to traditional North–South oriented greenhouses, East–West oriented greenhouses may require more sophisticated designs and higher-standard construction materials, resulting in increased initial investment costs.

High Technical Requirements: Ensuring a balanced greenhouse environment necessitates the use of advanced temperature and light control systems. This not only increases operational complexity but also requires farmers to possess certain technical knowledge and management skills.

Geographical and Climatic Limitations: The effectiveness of East–West oriented greenhouses is influenced by geographical location and climatic conditions. Adaptability and effectiveness may vary across different regions, necessitating local adjustments and optimizations.

Future Research and Practice Recommendations:

1. Focus on Regional Adaptability: Future research should concentrate on the performance of East–West oriented greenhouses under various climatic and geographical conditions. Extensive field testing and data analysis are needed to identify optimal design parameters and management strategies.
2. Development of Cost-Effective Technologies: Encourage and support research institutions and enterprises to develop more economical and practical greenhouse construction and management technologies. Finding ways to reduce costs will facilitate the widespread adoption of East–West oriented greenhouses.
3. Professional Training and Promotion: Organize professional training and promotional activities to improve farmers' technical proficiency and management capabilities. This will ensure that they can effectively leverage the advantages of East–West oriented greenhouses, thereby enhancing overall agricultural productivity and sustainability.

4. Government Support: Recommend that government departments provide policy and financial support to encourage agricultural producers to adopt East–West oriented greenhouse technology, particularly in resource-poor and ecologically fragile areas.

By addressing these challenges and following these recommendations, the advantages of East–West oriented greenhouses can be fully realized, contributing to the advancement of modern agriculture.

2.6 Substrate Bag Cultivation

With the rapid growth of the global population and the acceleration of urbanization, traditional agriculture faces numerous challenges, including land degradation, water scarcity, and environmental pollution. Substrate bag cultivation, a modern agricultural innovation, offers new possibilities for agricultural production in urban and marginal areas by using specially designed substrate bags and optimized cultivation media. This section delves into the basic concept of substrate bag cultivation, exploring its application background, design principles, environmental adaptability, and future development potential, aiming to provide an efficient and sustainable solution for global agricultural production.

Substrate bag cultivation technology originated from the need to address non-traditional agricultural environments, particularly in urban and suburban areas where land is limited or soil conditions are suboptimal. This technology uses specially designed containers (substrate bags) to hold soil substitutes such as coir, peat, and perlite, which can provide a good balance of water and air to support crop growth. Compared to traditional soil cultivation, substrate bag cultivation can reduce soil-borne diseases, improve water and nutrient use efficiency, and minimize environmental impact.

In terms of design principles, substrate bag cultivation takes into account the need for light, water, nutrients, and air circulation. Substrate bags are typically made of breathable and well-draining materials, such as non-woven fabric or specialized plastics, to adapt to different environmental conditions and crop needs. Additionally, the modular and mobile nature of this technology offers great flexibility in space utilization and crop rotation.

Regarding environmental adaptability, substrate bag cultivation technology performs exceptionally well. It can maintain stable production in harsh conditions such as drought, saline-alkali land, and low-fertility soils. By integrating drip irrigation and automated nutrient supply systems, precise agricultural management can be achieved. In urban environments, the use of substrate bags can support new agricultural forms such as rooftop greening, vertical farming, and indoor agriculture, increasing urban green spaces while providing fresh agricultural products.

Looking ahead, the progress in biotechnology, materials science, and automation technology is expected to further optimize substrate bag cultivation. Research should focus on developing more environmentally friendly and cost-effective substrate materials and enhancing system automation levels to reduce labor costs. Additionally, field trials and data analysis should be conducted to further verify and improve the performance of substrate bag cultivation under various climatic and geographical conditions, aiding in the promotion of this technology to a broader range of global regions. In conclusion, substrate bag cultivation technology, as an efficient and eco-friendly modern agricultural solution, holds great promise for global agricultural production. Through technological innovation and policy support, its wider application is anticipated, contributing to global food security and sustainable development goals.

2.6.1 Concept and Application Background of Substrate Bag Cultivation

With the growth of the global population and the rapid advancement of urbanization, traditional agriculture faces numerous challenges, including limited land resources, soil degradation, water scarcity, and environmental pollution. In this context, substrate bag cultivation technology has emerged as a sustainable agricultural solution, particularly suitable for areas with limited land or fragile ecological environments.

Substrate bag cultivation is a technique that involves the use of containers (usually plastic bags or fabric bags) filled with soilless growing media for plant cultivation. This media may include materials such as coir, peat, perlite, and vermiculite. These materials not only have good water retention and aeration properties but also provide the necessary nutrients to support plant growth. The use of substrate bags eliminates the dependency on soil for cultivation, significantly expanding the possibilities for agricultural activities, especially in non-traditional agricultural environments.

Substrate bag cultivation technology was initially developed to address specific issues such as soil salinization, frequent pest infestations, and insufficient soil fertility. As the technology matured and its range of applications expanded, it has come to be widely used in various agricultural production scenarios, including:

Urban Agriculture: In urban and suburban areas where land resources are scarce and costly, traditional ground farming methods are highly restricted. Substrate bag cultivation technology can be implemented on rooftops, balconies, and even indoor environments, greatly promoting the development of urban agriculture and vertical farming.

Agricultural Production in Constrained Areas: In deserts, rocky terrains, and other regions with harsh natural conditions, the land is almost unusable for agricultural production. Substrate bags provide a feasible method for conducting agricultural activities in these areas. By precisely controlling the supply of water and nutrients, efficient plant growth can be achieved.

Post-Disaster Reconstruction: Following natural disasters such as floods and earthquakes, land is often eroded or polluted, making it difficult to immediately resume agricultural production. Substrate bag cultivation can serve as a rapidly deployable temporary agricultural production method, helping to restore food supply while facilitating ecological restoration of the land.

Sustainable Agriculture: With increasing awareness of environmental protection, sustainable agriculture is receiving more attention. Substrate bag cultivation, with its advantages of water conservation, reduced use of chemical fertilizers and pesticides, and lower risk of land degradation, has become an important tool for achieving sustainable agricultural goals.

The advantages of substrate bag cultivation lie in its high flexibility, strong controllability, and good environmental adaptability. It can reduce the occurrence of soil-borne diseases and, through precise water and nutrient management, improve resource use efficiency. Additionally, the lightweight nature of substrate bags makes them easy to install and move in various environments, providing unprecedented flexibility and convenience for agricultural production.

2.6.2 Design Principles of Substrate Bag Cultivation

Substrate bag cultivation is an efficient agricultural technology that uses specially designed containers (substrate bags) for crop cultivation, replacing traditional soil farming methods. This method is not only suitable for resource-limited environments such as urban rooftops, deserts, or rocky areas, but also demonstrates significant advantages in controlling crop growth conditions and improving production efficiency. This section will explore the design principles of substrate bag cultivation in detail, including substrate selection, bag material characteristics, and the design of water and nutrient management systems (Fig. 2.16).

A core element of substrate bag cultivation is the selection of substrates. The ideal substrate should have good water retention, aeration, appropriate buffering capacity, and sufficient mechanical strength. Common substrate materials include:

- Coir: Derived from coconut husk fibers, this renewable resource has excellent water and air regulation capabilities and a relatively low pH.
- Peat: Known for its excellent water retention and buffering capacity, peat is often used to adjust the substrate's pH and increase organic matter content.
- Perlite and Vermiculite: These inorganic materials provide good drainage and aeration, helping root development and preventing root diseases.
- Vermiculite: Noted for its good water retention and nutrient-holding capacity, it also provides a loose root environment.

The choice of substrate should be adjusted based on the specific needs of the crops, local climate conditions, and economic feasibility. The ideal substrate should meet all the physiological needs of the plants, including support, water, nutrients, and oxygen.

Fig. 2.16 Vertical substrate bag cultivation model

The materials of substrate bags need to have sufficient strength and durability while allowing the passage of air and moisture. The choice of material directly affects the quality of the root zone environment and plant growth conditions. Common materials include:

Non-woven Fabric: Good breathability, able to maintain moisture and aeration of the substrate, suitable for various climate conditions.

High-Density Polyethylene (HDPE): Chemically resistant and hydrolysis-resistant, suitable for long-term outdoor use, but with poor breathability that requires appropriate design to ensure root ventilation.

Polypropylene (PP) Fabric: Lightweight and moderately priced, often used for disposable or short-term cultivation projects.

Water and nutrient management are critical parts of the substrate bag cultivation system design. A reasonable water and fertilizer management system can not only ensure the healthy growth of crops but also improve resource use efficiency and reduce environmental impact. Design principles include:

Drip Irrigation System: Precisely controls the supply of water and nutrients, maximizing water and fertilizer use efficiency. The drip irrigation system can automatically adjust water and nutrient supply based on the crop's growth stage and climate conditions.

Drainage Design: A good drainage system is key to preventing root diseases and waterlogging. The bottom and sides of the substrate bag should be designed with appropriate drainage holes to prevent excessive water accumulation (Fig. 2.17).

Fig. 2.17 Horizontal substrate bag cultivation model

The design principles of substrate bag cultivation focus on providing a controlled environment by finely tuning the substrate, bag materials, water and fertilizer systems, and structural layout to optimize plant growth conditions. This technology not only offers greater flexibility and sustainability for agricultural production but also opens up new possibilities for urban agriculture and resource-limited areas. With advancements in technology and increasing environmental pressures, substrate bag cultivation technology is expected to play an increasingly important role in global agricultural production.

2.6.3 Environmental Adaptability of Substrate Bag Cultivation

Due to its unique cultivation method, substrate bag cultivation technology can maintain the stability and productivity of crop growth under various climatic conditions. Whether in tropical, temperate, or cold regions, the use of substrate bags can adapt to different climate needs by adjusting the substrate composition and drip irrigation system:

Tropical and Subtropical Regions: In these areas, high temperatures and humidity are the main challenges for crop growth. Substrate bags can reduce the occurrence of root diseases by using substrates with good aeration and drainage properties, such as coir or perlite. Additionally, the drip irrigation system can accurately control the supply of water and nutrients, preventing over-irrigation and nutrient loss.

Temperate Regions: Temperate climates have distinct seasonal changes. Substrate bag cultivation can extend the growing season by incorporating insulation measures, such as using thermal cover materials. Moreover, appropriate substrate selection can help maintain root temperature during cooler seasons, promoting healthy growth.

Cold Regions: In cold climates, substrate bag cultivation can ensure crop growth during winter by incorporating heating systems, such as bottom heating pads or warm irrigation water. This technology reduces the risk of damage caused by frost.

Another advantage of substrate bag cultivation technology is its wide adaptability, supporting various types of crops from leafy vegetables to fruit trees:

Leafy Vegetables (e.g., Lettuce, Spinach): These environment-sensitive crops benefit from the stable environment provided by substrate bags, ensuring consistent supply of light, water, and nutrients.

Fruit-Bearing Crops (e.g., Tomatoes, Peppers): The controlled environment of substrate bags helps manage and optimize the specific conditions required for fruit growth, such as precise control of nutrient ratios and water cycles, thereby increasing fruit size and yield.

Medicinal Plants: These crops often require special nutrient management and growth environments. Substrate bags allow for specific substrate formulations and adjustable nutrient supply plans to meet these needs.

Substrate bag cultivation is not only adaptable to natural environments but is also suitable for application in urban environments, providing a sustainable urban agriculture solution:

Efficient Use of Urban Space: Substrate bag cultivation can be conducted in limited spaces such as rooftops, vacant lots, or even indoors, maximizing space utilization through vertical farming.

Environmental Friendliness and Community Engagement: Substrate bag cultivation promotes urban greening and can serve as an educational tool and focal point for community activities, raising awareness about sustainable agriculture among citizens.

2.6.4 Conclusion

As an innovation in the field of modern agriculture, substrate bag cultivation technology demonstrates strong adaptability to various environmental conditions and holds significant potential for widespread application. This soilless cultivation method effectively addresses issues such as land resource scarcity, soil degradation, and water shortages. Moreover, its inherent flexibility and sustainability offer new solutions for urban agriculture and farming in resource-limited areas.

Substrate bag cultivation technology utilizes specially designed containers and specific substrate combinations, enabling agricultural production on non-traditional lands. The introduction of this technology not only enhances the efficiency and quality of crop production but also reduces agriculture's environmental impact. The design of substrate bags considers the needs for light, moisture, nutrients, and air circulation, ensuring that crops grow under optimal conditions. Additionally, with drip irrigation and other precise irrigation systems, efficient use of water and fertilizers is achieved.

Furthermore, due to its modularity and mobility, substrate bag cultivation is particularly suited for urban and vertical farming environments. It provides urban residents with opportunities to engage in green activities and offers a local option for food production, thereby reducing the carbon footprint and food waste associated with transportation.

Looking ahead, as technology continues to improve and costs decrease, substrate bag cultivation is expected to see more widespread application globally. To achieve this goal, it is recommended to increase investment in research and development, optimize substrate formulations, improve bag materials, and integrate smart agricultural technologies such as automated water and fertilizer systems and climate control equipment. These steps will enhance the overall performance and economic viability of the system.

2.7 Elevated Cultivation Systems

Elevated cultivation systems, as a modern agricultural technology, optimize control over crop growth environments and maximize space utilization by raising the crop cultivation bases. This technology shows immense application potential in urban agriculture, agriculture in confined spaces, and regions with adverse growing conditions. This section will explore the concept and application background of elevated cultivation systems, design principles, environmental adaptability, and their contribution to future agricultural development.

2.7.1 Concept and Application Background of Elevated Cultivation Systems

Elevated cultivation systems generally refer to methods of crop cultivation through platforms or frameworks elevated above ground level. These systems can be constructed from various materials such as metal, wood, plastic, or composite materials. They are highly flexible in design and can be customized according to specific environmental and crop needs. The cultivation bases can be filled with soil or alternative soilless media like coir or perlite.

With the accelerated process of global urbanization, land resources available for agriculture are increasingly scarce. In urban areas, especially densely populated metropolitan regions, traditional ground-based farming can no longer meet the demand for green food. Elevated cultivation systems can effectively utilize unused urban spaces such as rooftops, vacant lots, and even walls for vertical farming, optimizing space utilization and enhancing the feasibility of agricultural production.

Globally, soil degradation caused by over-farming, chemical pollution, and deteriorating natural conditions has become a significant barrier to agricultural development. Elevated cultivation systems reduce dependence on soil, offering a new solution for sustainable agriculture. In arid and semi-arid regions, effective water management is crucial for agriculture. Elevated systems are often equipped with efficient irrigation systems such as drip irrigation, which can precisely control the supply of water and nutrients, reducing water waste and increasing water efficiency in crop production.

Elevated cultivation systems contribute to environmental protection by reducing ground disturbance and chemical inputs. They support the implementation of organic farming by reducing the use of fertilizers and pesticides, thereby helping to maintain biodiversity.

In multiple countries and regions worldwide, elevated cultivation systems have been successfully implemented, demonstrating significant economic and environmental benefits. For example, in Japan and South Korea, elevated cultivation systems are widely used in urban agriculture, improving the food supply chain for urban residents and increasing green living spaces. In arid regions of Africa and the Middle East, these systems help local farmers overcome harsh natural conditions, achieving stable crop production.

2.7.2 Design Principles of Elevated Cultivation Systems

As an advanced agricultural technology, elevated cultivation systems aim to improve spatial efficiency, enhance crop growth conditions, and promote environmental sustainability. The design of these systems integrates multiple considerations, from structural stability to irrigation efficiency, with every detail aimed at optimizing crop production and management. The following is a detailed analysis of the design principles of elevated cultivation systems (Fig. 2.18).

Material Selection: The structures of elevated cultivation systems typically use corrosion-resistant and weather-resistant materials such as galvanized steel, stainless steel, aluminum, or durable wood (e.g., cedar or pressure-treated wood). These materials ensure structural stability and long-term durability while also considering environmental friendliness and sustainability.

Structural Stability: The design must ensure that the structure can withstand the load from crops, substrates, irrigation systems, and climatic conditions (e.g., wind loads, snow loads). Additionally, the height, width, and length of the elevated system need to be optimized based on the type of crops and cultivation density to support normal crop growth and facilitate daily management.

Accessibility and Ergonomics: The design of the elevated system should allow farmers or users to easily access the crops for planting, maintenance, and harvesting.

Fig. 2.18 Elevated cultivation planting model diagram

The height and width design should follow ergonomic principles to reduce labor intensity and improve work efficiency.

Irrigation System Selection: Common irrigation methods for elevated cultivation systems include drip irrigation and micro-sprinkler irrigation. These methods can precisely control the supply of water and nutrients, reducing water waste. The system design should ensure uniform water distribution, avoiding localized over-watering or under-watering, which could affect crop growth.

Automation and Monitoring: Modern elevated systems are often equipped with sensors and automatic control technologies to monitor soil moisture, nutrient levels, and environmental variables (such as temperature and humidity) in real-time. This information can be used to automatically adjust irrigation and fertilization plans, optimizing the crop growth environment.

Drainage Design: Elevated cultivation beds must have effective drainage systems to prevent water accumulation, which could cause root diseases. The design should include drainage holes at the bottom and sides of the bed, as well as potential drainage ditches or collection systems to ensure prompt removal of excess moisture.

Substrate Selection: The substrate needs to provide good support, water retention, aeration, and nutrient-holding capacity. Commonly selected substrates include mixtures of peat, coir, perlite, and vermiculite. The pH, salinity, and organic matter content of the substrate should be adjusted according to the needs of the crops.

Root Zone Environment Optimization: The health of the root zone is crucial for crop growth. When designing elevated cultivation systems, sufficient space for root development should be ensured. Additionally, substrate and irrigation management should maintain appropriate oxygen and moisture conditions for the roots.

2.7.3 Environmental Adaptability of Elevated Cultivation Systems

As a flexible agricultural technology, elevated cultivation systems allow for crop production under diverse environmental conditions through their unique design. These systems can overcome various land and climate constraints, displaying exceptional adaptability from urban rooftops to desert areas, from temperate zones to tropical regions. The following section discusses in detail the environmental adaptability of elevated cultivation systems, including their ability to be applied under different geographical and climatic conditions.

The design of elevated cultivation systems allows them to operate under a wide range of global climatic conditions, including but not limited to the following:

Temperate Climate: In temperate climates, elevated systems can adapt to seasonal changes by adjusting the covering materials. For example, using insulating materials like agricultural films during winter can reduce heat loss and extend the growing season. In summer, appropriate shading measures can prevent excessive temperatures from stressing the plants.

Tropical and Subtropical Climates: In these regions, elevated cultivation systems address the challenges of high temperatures and humidity by optimizing irrigation and moisture management. Drip irrigation systems can precisely control water supply, avoiding root problems caused by excess moisture while ensuring adequate plant hydration under high heat conditions.

Arid and Semi-Arid Areas: Elevated systems are particularly suitable for water-limited arid regions because they can be equipped with efficient water resource management systems such as drip and mist irrigation. These systems significantly reduce water evaporation and increase water use efficiency.

The design of elevated cultivation systems is also suitable for different geographical environments, demonstrating their advantages in the following situations:

Urban Environments: Due to their high spatial efficiency, elevated systems are particularly suited for space-limited urban environments. They can be set up on rooftops, balconies, and even indoor spaces, making use of previously underutilized areas. This not only maximizes space utilization but also allows urban residents to participate directly in food production, enhancing food security and reducing food miles.

Restricted and Marginal Areas: In areas with poor land quality or contaminated soil, such as industrial wastelands and coastal saline-alkaline soils, traditional ground-based farming methods are limited. Elevated cultivation systems can effectively isolate from ground conditions, allowing agricultural production in almost all types of geographical environments.

While improving yield and production efficiency, elevated cultivation systems also focus on environmental protection:

Reducing Soil Erosion and Pollution: Since elevated systems do not rely on soil, they reduce the potential for soil erosion and chemical pollution. This is especially important for protecting the ecological environment in industrial areas with a long history of chemical pollution.

Sustainable Resource Management: The integrated water and fertilizer technology and precise irrigation strategies in system design help rationalize resource utilization, reduce the consumption of water and fertilizers, and enhance the overall sustainability of agriculture.

2.7.4 Summary of Elevated Cultivation Systems

As an essential component of modern agricultural technology, elevated cultivation systems effectively address issues such as land resource scarcity, environmental degradation, and the need for increased production efficiency through their innovative design concepts and methods. These systems not only optimize space utilization and enhance crop growth conditions but also significantly improve the environmental adaptability and sustainability of agricultural production.

Elevated cultivation systems demonstrate their flexibility and efficiency under various climatic and geographical conditions. Whether in temperate, tropical, or arid regions, these systems ensure optimal crop growth through precisely controlled irrigation and nutrient supply systems, while minimizing resource waste. In urban environments, they showcase their immense potential under space-limited conditions, utilizing rooftops, vacant lots, and even walls for agricultural production, thereby directly increasing urban green space and improving residents' quality of life.

Furthermore, elevated cultivation systems play an important role in environmental protection. By reducing reliance on soil, these systems effectively avoid the risks of soil erosion and chemical pollution. They support sustainable agricultural practices such as organic farming and ecological farming, contributing to ecological conservation.

2.8 Summary

This study delved into different cultivation models using yellow sand substrate in desert and Gobi areas, focusing on the adaptability and effectiveness of north–south and east–west oriented greenhouse cultivation models, substrate bag planting, and elevated cultivation systems. The study revealed that the physical properties of yellow sand, such as high permeability and good drainage, are beneficial for reducing root water retention and disease occurrence. However, its poor nutrient retention capacity requires additional cultivation techniques and management strategies, such as adding organic matter and nutrients to enhance its fertility.

Comparative analysis showed that different cultivation orientations significantly affect crop growth. For instance, the north–south oriented greenhouse model can maximize sunlight utilization during winter, while the east–west orientation improves photosynthetic efficiency of plant leaves during spring and autumn. Additionally, substrate bag planting and elevated cultivation systems exhibit unique advantages in space-limited or complex terrain areas of the Gobi, such as high flexibility, easy management, and disease control, offering new possibilities for agricultural use of marginal lands.

Experimental results further validated the performance and adaptability of different cultivation models under extreme climatic conditions. North–south oriented greenhouses, by receiving consistent sunlight throughout most of the day, particularly reduce midday sun exposure during summer, effectively alleviating high-temperature stress on crops. This helps extend the growing season and improve crop yield and quality. East–west oriented greenhouses, under low solar angles in spring and autumn, allow plants to receive sunlight more evenly, reducing the risk of photosynthetic saturation or inhibition, thus optimizing light utilization. Substrate bag planting, with its independent cultivation units and controlled environment conditions, effectively avoids soil succession barriers and significantly improves water and nutrient use efficiency and crop growth speed through precise water and nutrient management systems. Elevated cultivation systems, by raising the planting beds, reduce the impact of ground heat radiation on crops, facilitate crop management and pest control, and are especially suitable for urban and constrained space agricultural applications.

The study's conclusions emphasize the importance of scientific cultivation techniques and soil improvement methods in promoting agricultural development in extreme environments. Through continuous technological innovation and optimization of cultivation methods, the agricultural production potential of desert and Gobi regions can be fully realized, contributing to global food security and ecological restoration. Future research should focus more on the practical performance of cultivation models under different climatic and geographical conditions and how to further improve the nutrient and water retention capacity of yellow sand substrates. Additionally, adopting modern agricultural technologies such as automation and precision agriculture tools will be key to enhancing the sustainability and efficiency of desert agriculture. Through international cooperation and policy support, these cultivation techniques can be promoted globally, particularly in resource-poor and ecologically fragile regions.

Chapter 3
Design and Testing of a Self-propelled Trenching and Mulching Machine for Desert Gobi Sand Substrate

Yalong Song, Jianfei Xing, Long Wang, Jiahui Xu, Xufeng Wang, and Jianing Long

3.1 Introduction

Desertification poses a serious threat to human development and sustainability. In the northwestern regions of China, harsh land conditions dominated by deserts, Gobi, and saline-alkali land significantly constrain high-quality agricultural development. In recent years, China has promoted facility agriculture in non-arable areas to address the shortage of arable land resources. In facility agriculture, soilless cultivation techniques are predominantly used, and the embedded substrate cultivation model has been widely applied in desertified areas. This model features functions such as evaporation suppression with plastic film, water retention and conservation, rainwater harvesting, and drought resistance. It also effectively prevents issues such as

Y. Song · J. Xing (✉) · L. Wang · J. Xu · X. Wang (✉)
College of Mechanical and Electrical Engineering, Tarim University, Alar 843300, China
e-mail: 591101353@qq.com

X. Wang
e-mail: wxfwyq@126.com

Modern Agricultural Engineering Key Laboratory at Universities of Education Department of Xinjiang Uygur Autonomous Region, Alar 843300, China

Xinjiang Production and Construction Corps (XPCC) Key Laboratory of Utilization and Equipment of Special Agricultural and Forestry Products in Southern Xinjiang, Alar 843300, China

L. Wang
e-mail: 120140002@taru.edu.cn

J. Long (✉)
Key Laboratory of Agricultural Information Acquisition Technology, Ministry of Agriculture and Rural Affairs, China Agricultural University, Beijing 100083, China
e-mail: 614020890@qq.com

C. Hu et al. (eds.), *New Technologies Applied in Gobi Desert Facility Agriculture*, Smart Agriculture 11, https://doi.org/10.1007/978-981-96-1031-0_3

low-temperature stress and soil continuous cropping obstacles, thereby enhancing the yield of fruits and vegetables. In solar greenhouses, challenges such as limited space, sandy soils, large traditional trenching and mulching machines, low trenching quality, and suboptimal soil throwing performance are prevalent. Therefore, developing a trenching device suitable for the sandy substrate cultivation model in solar greenhouses is of great significance.

Countries such as Germany, the United Kingdom, the United States, and Canada have conducted early and extensive research in the field of trenching and mulching equipment. Japan started related research as early as 1948 and is one of the first countries in the world to promote trenching and mulching sowing. Its large-scale field machinery, such as ridge mulching machines and mulching seeders, has been widely applied. In 2023, Hamroqul Ravshanov and colleagues developed a combination trenching blade that, through the optimization of its height, length, wing angle, and installation angle, can achieve a 180° soil layer turnover, providing a favorable environment for crop root systems. However, due to the loose nature of sandy soils, a 180° turnover can lead to trench instability, affecting planting effectiveness. Sukhbir Singh and colleagues designed the IISR deep trenching blade, which can be adjusted according to speed and depth to control soil disturbance and traction demand. However, sandy soils may cause significant fluctuations in traction demand during operation, impacting consistency and operational stability. James Barr and colleagues designed a curved-leg trenching blade with a streamlined design and optimized thickness to minimize soil disturbance. However, further experimentation and design optimization are needed to address the unique physical properties of sandy soils. K. Ravshanov and colleagues designed a multifunctional trenching machine that can form irrigation channels along the symmetrical axis of the planting zone, optimizing water resource usage and distribution to improve irrigation efficiency. However, in sandy environments, the irrigation channels are unstable and prone to collapse, requiring enhanced designs to ensure channel stability. Neil B. McLaughlin and colleagues compared two different trenching blades to provide options for various soil and planting conditions. Their research offers insights and assistance in designing trenching blades suitable for sandy cultivation, requiring structural optimization to adapt to local soil conditions, reduce resistance, enhance operational efficiency, and ensure stability and effectiveness in loose sandy soils. Despite these advancements, the differing soil properties abroad often lead to the development of high-power, wide-span machinery with high technical requirements, making them relatively difficult for farmers to operate. Therefore, such equipment is not suitable for application in the severely desertified regions of northwest China.

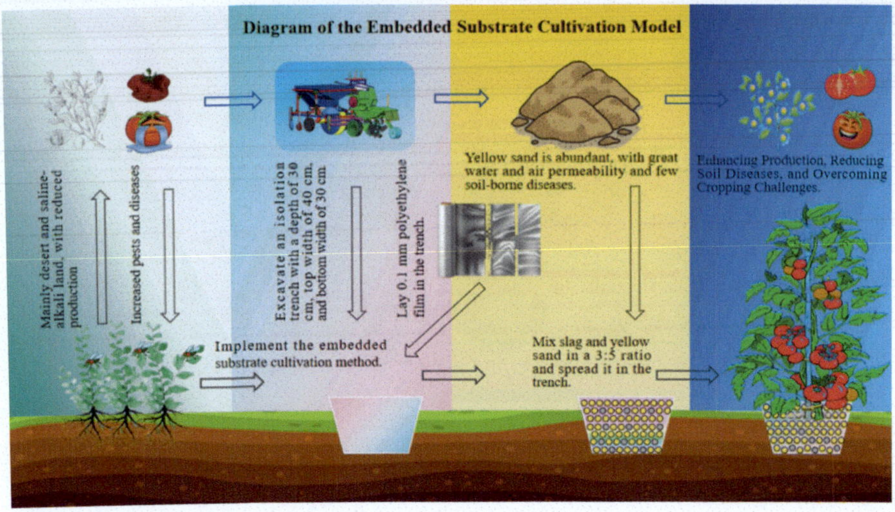

Fig. 3.1 In-situ substrate cultivation model diagram

3.2 Materials and Methods

3.2.1 Agronomic Requirements

To enhance production efficiency and mitigate secondary soil salinization, an in-situ substrate cultivation model is employed. This model involves the excavation of isolation trenches (planting furrows) with a depth of 30 cm, a surface width of 40 cm, and a bottom width of 30 cm. The distance between adjacent planting furrows is 80 cm. A polyethylene film, approximately 0.1 mm thick, is laid within the planting furrows to create a physical barrier, reducing pest and disease attacks on the crops. Subsequently, slag and yellow sand are thoroughly mixed at a mass ratio of 3:5 and evenly spread within the planting furrows. This mixture improves soil aeration and drainage, providing a conducive environment for root growth, preventing secondary soil salinization, and enhancing land use efficiency. This setup is illustrated in Fig. 3.1.

3.2.2 Overall Structure of the Equipment

The trenching and film-covering integrated machine mainly consists of the following components: guiding and depth-limiting mechanism, frame, soil-retaining and back-flow device, drive wheel, gearbox, gasoline tank, film-covering mechanism, trapezoidal pressing mechanism, and substrate laying mechanism. A schematic diagram of this structure is shown in Fig. 3.2.

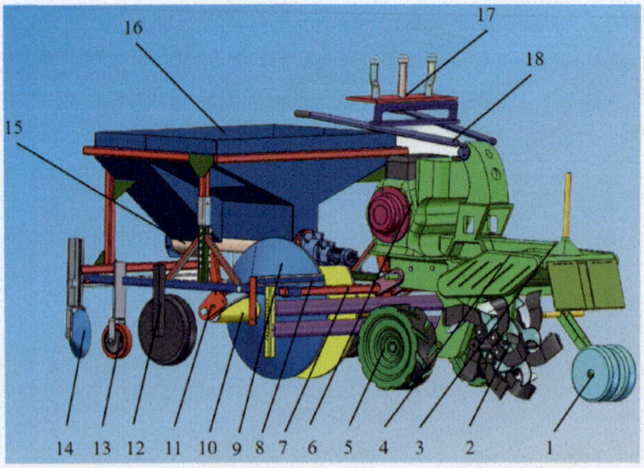

1. Guide and depth limiting device 2. Trench opener 3. Upper retaining plate 4. Side retaining plate 5. Driving wheel 6. Engine 7. Frame 8. Film frame 9. Suppression device 10. Suppression roller 11. Multi-link regulating mechanism 12. Walking wheel 13. Film pressing wheel 14. Earth covering disc 15. Winch 16. Matrix box 17. Gear adjustment lever 18. Joystick

Fig. 3.2 Complete structure of the trenching and film-covering integrated machine

3.2.3 Working Principle

The guiding and depth-limiting device consists of a support frame and guiding wheels, allowing the trenching depth to be controlled by adjusting the height of the guiding wheels. The control device includes a handheld control frame and a gear adjustment system. During operation, the pulley splits the power into two paths: one path transmits power to the gearbox to drive the trenching blade shaft, while the other path propels the machine forward. The rotary tiller drives the trenching blade to rotate counterclockwise, with the helically staggered trenching blades crushing the soil and throwing it to both sides of the trench. The soil-retaining plate, fixed above the trencher, ensures that the soil is evenly distributed along the edges of the trench, preventing soil from falling back and maintaining the trench shape and dimensions. The film-covering device is located at the front of the frame, completing the film-covering operation as the machine advances. The walking pressing device simultaneously presses the film and secures the planting trench. The substrate box, welded above the frame, evenly distributes substrate into the planting trench during operation, completing the substrate laying task.

3.2.4 *Design of Key Components*

Design of the Throwing and Cutting Combination Trencher

The narrow space inside solar greenhouses limits the use of large agricultural equipment. Traditional plowshare and chain blade trenching devices have high tillage resistance and demand significant engine power. In contrast, rotary tiller trenching devices have a simple structure, excellent soil cutting and crushing performance, stable operation, and low power consumption, making them suitable for the needs of facility agriculture. Currently, disc-type trenchers typically use universally spaced rotary tiller blades arranged equidistantly along the circumference (as shown in Fig. 3.3). Although this design has good soil cutting performance, its soil throwing performance is poor, resulting in insufficient trench width, severe soil backfill, and inability to meet the operational requirements of solar greenhouses.

To address this issue, and considering the characteristics of sandy soil, sand particles, and gravel prevalent in the northwest region of China, this research designed an improved throwing and cutting combination trencher. This device uses staggered rotary tiller blades for cutting soil and adds throwing plates to improve soil throwing performance (as shown in Fig. 3.4).

Optimization of Key Parameters

To meet the agronomic requirements of trench depths controlled around 300 mm, this design adjusts key parameters of the traditional IT245 rotary tiller blades. To enhance the soil-crushing capability of the throwing and cutting combination trencher, increasing the height of the leading edge of the blade's cutting surface can improve its soil-fragmentation performance. When the height of the leading edge of

Fig. 3.3 Schematic diagram of the single soil-cutting trencher structure

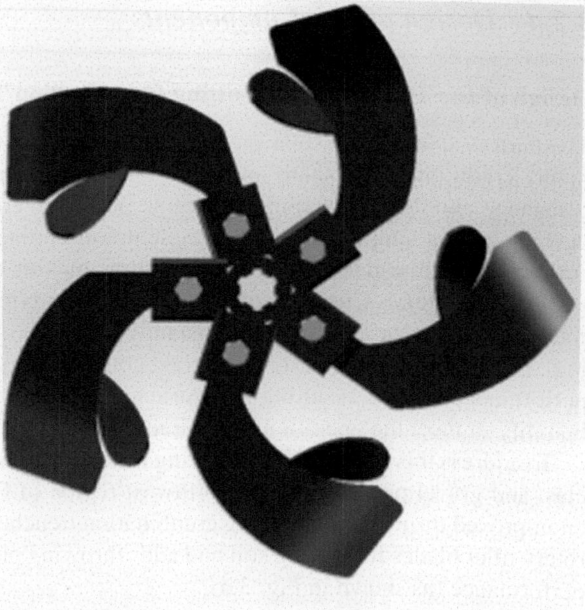

Fig. 3.4 Schematic diagram of the optimized throwing and cutting combination trencher structure

the cutting surface is 65 mm, both soil fragmentation rate and power consumption are optimal. Therefore, the height of the leading edge of the throwing blade is increased from 40 to 65 mm, while other parameters remain unchanged. The improved blade structure is shown in Figs. 3.5 and 3.6.

Taking the rotation center of the trencher blade shaft as the origin of the coordinates, and considering the forward direction of the machine as the x-axis and the

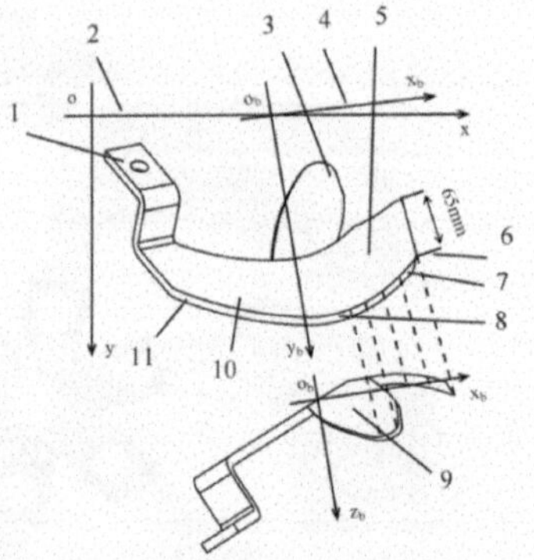

1. Tool handle 2. Side view coordinates of side cut 3. Chipper 4. Side view coordinates of front position 5. Cutting surface 6. Face height 7. Side projection 8. Cutting edge 9. Chipper 10. Side section 11. Side cutting edge

Fig. 3.5 Schematic diagram of the end height of the positive section of the soil throwing knife

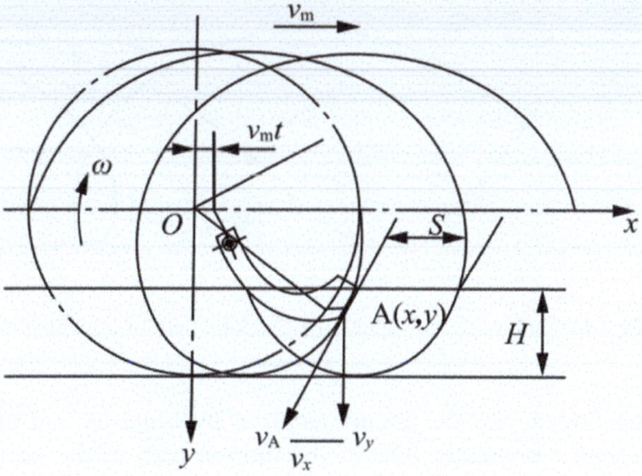

Fig. 3.6 Motion diagram of soil cutter

positive direction of the y-axis as vertically upward, a Cartesian coordinate system is established. The end point of the blade coincides with the y-axis, and its motion equations are given by the following formulas (3.1)–(3.3):

$$x = R\cos\omega t + v_m t \tag{3.1}$$

$$y = R\sin\omega t \tag{3.2}$$

$$\lambda = \frac{R\omega}{v_m} \tag{3.3}$$

The absolute velocity v of the end point of the grooving blade is shown in formula (3.4):

$$v = \sqrt{v_x^2 + v_y^2} = \sqrt{v_o^2 + R^2 - 2v_m R\omega\sin\omega t} \tag{3.4}$$

where: R—rotation radius of rotary plow knife, mm;
ω—rotational angular velocity of rotary tilting knife, rad/s;
v_m—forward speed of the tool, m/s;
λ—rotary tillage speed ratio;
t—Time.

Rotary tillage speed ratio has an important influence on the operating quality of the trenching tool, and the influence on the trajectory of the trenching tool is shown in Fig. 3.7. When $\lambda = 1$, the motion path of the end point of the trenching tool is cycloidal, and the end point of the trenching tool cannot cut the soil backward and cannot break the soil. When $\lambda < 1$, the horizontal component speed of the end point of the trenching knife is always the same as the direction of the rototiller, and the movement trajectory is a short cycloid, the trenching knife cannot cut the soil

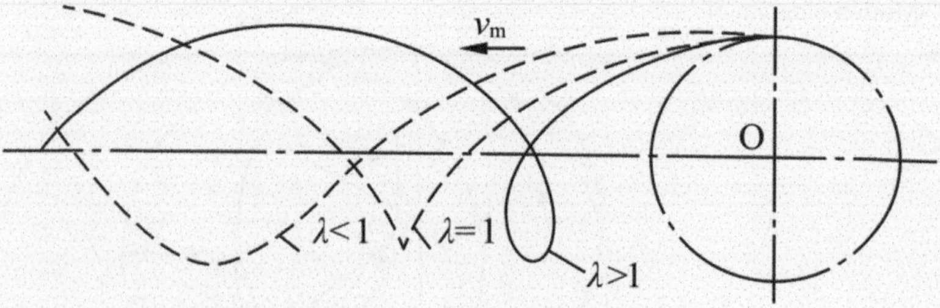

Fig. 3.7 Influence of λ value on the motion path of trenching tool

backward, and the phenomenon of pushing the soil forward cannot work properly. When λ is greater than 1, the motion path of the trenching tool is a trochoid, and the horizontal component velocity of the trenching tool endpoint is opposite to the direction of the machine, and the trenching tool can cut soil backward to complete the trenching operation and ensure the quality of the trenching operation.

Cutter diameter affects the trenching power consumption. The calculation formula is shown in (3.5):

$$D = (1.2\text{–}1.4)H_1 \tag{3.5}$$

$$D = 360\text{–}420\,\text{mm} \tag{3.6}$$

Formula: H_1—Trench depth (300 mm);

D—Blade disk diameter.

According to the agricultural requirements for facility agriculture, the trenching depth needs to reach 300 mm. To ensure that the combined cutting and throwing trencher blade can uniformly cut the soil and throw it onto the trench surface while reducing power consumption, the designed blade disk rotation diameter is 400 mm. The cutting interval affects the soil fragmentation effect and the stability of trenching and ridge formation. Different soils and crops require different cutting intervals. Generally, the larger the interval, the lower the soil fragmentation rate. For soils with low viscosity and high water content, a smaller interval should be chosen; for soils with high viscosity and low water content, a larger interval should be chosen. The specific formulas are shown below.

$$S = v_m t_s = v_m \frac{2\pi R}{\lambda z} = \frac{60 v_m}{zn} \tag{3.7}$$

where: S—Cutting interval (soil cutting pitch);

Z—Number of tilling blades installed, one;

n—Blade shaft speed (revolutions per minute, r/min);

v_m—Forward speed of the machine (meters per second, m/s);

t_s—Working rotation time of the blade disk (seconds, s);

R—Blade disk radius (millimeters, mm);
ω—Blade disk angular velocity (radians per second, rad/s);
λ—Tilling speed ratio.

It can be seen from the above equations that adjusting the forward speed of the machine, the number of blades, and the rotation speed of the blade disk can all change the cutting interval. Using a small hand-held rotary tiller as the driver, its forward speed is 0.5 m/s. In this article, the cutting interval for trenching and ridging blades is defined as 15–50 mm.

Analysis of Soil Throwing Process by the Soil Throwing Blade

The combined cutting and throwing trencher blade can laterally throw soil to a designated position primarily due to its soil throwing blade structure. According to the agronomic requirements of the embedded substrate cultivation mode, soil needs to be evenly thrown to the edge of the trench, 300 mm away, during trenching to facilitate subsequent film mulching operations. Therefore, it is necessary to design the inclination angle and width of the soil throwing blade. Since the soil in the northwest region mainly consists of sandy soil with typical discrete characteristics, soil particles can be regarded as particles. When the soil leaves the soil throwing blade, its trajectory follows a parabolic path, as shown in Fig. 3.8.

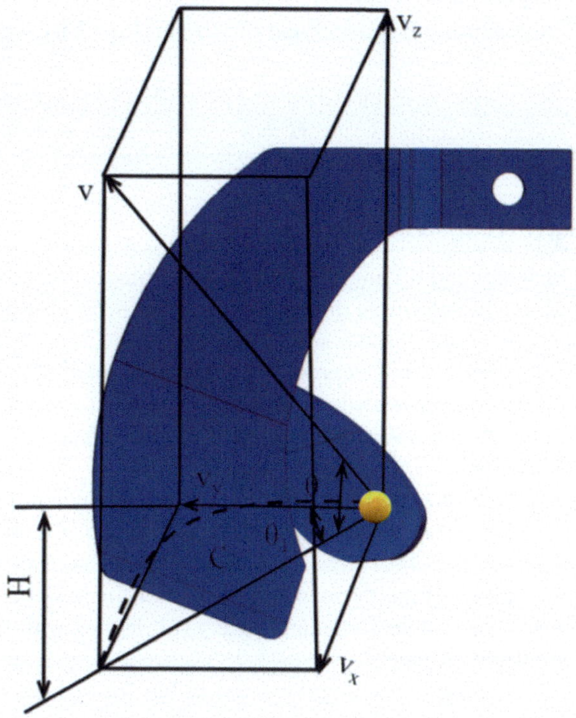

Fig. 3.8 Analysis of the process of soil dumping

The distance of the soil thrown from the soil throwing blade can be calculated using the following formula:

$$C = \frac{v^2 \sin(2\theta)}{g} \tag{3.8}$$

$$v^2 = v_x^2 + v_y^2 + v_z^2 \tag{3.9}$$

where: C—throwing distance, m;
v—initial velocity when thrown, m/s;
θ—angle between initial velocity and the ground, (°);
g—acceleration due to gravity, 9.8 m^2/s;
v_x, v_y, v_z—The component of the initial (absolute) velocity v in the three-dimensional rectangular coordinate system of space, m/s.

As shown in Fig. 3.7, the horizontal throwing distance is:

$$H = C \sin\theta_1 \tag{3.10}$$

where: θ_1—angle between the projection of v on the ground and V_x, (°);
H—horizontal throwing distance, m;
and θ_1 satisfies $\tan\theta_1 = v_y/v_x$.

Substituting Eqs. (3.9) and (3.10) into Eq. (3.8), we get:

$$H = \frac{\left(v_x^2 + v_y^2 + v_z^2\right)\sin(2\theta)v_y}{g\sqrt{v_x^2 + v_y^2}} \tag{3.11}$$

Force Analysis of Throwing Blade

The width and tilt angle of the throwing blade affect the velocity components v_x, v_y and, v_z. To determine the appropriate parameters for the throwing blade, a dynamic analysis of the thrown soil is necessary. In Fig. 3.9, a coordinate system oxyz is established, where O represents the rotation center of the throwing blade. The positive Y-axis is aligned with the forward direction of the machinery, the X-axis is horizontal, and the Z-axis is vertically upward. The front face of the throwing blade handle is in the yoz plane.

To facilitate the representation of the parameters related to the throwing blade, the intersection of the long and short edges of the throwing blade is taken as the origin o_1. The short edge is defined as the x_1-axis and the long edge as the y_1-axis. In the plane of the throwing blade, a coordinate system $x_1o_1y_1$ is established. This coordinate system forms an angle γ (tilt angle) with the yoz plane and an angle β with the xoy plane.

According to the angular relationship formed between the soil throwing sheet, yoz and xoy, v_x, v_y and, v_z are expressed as:

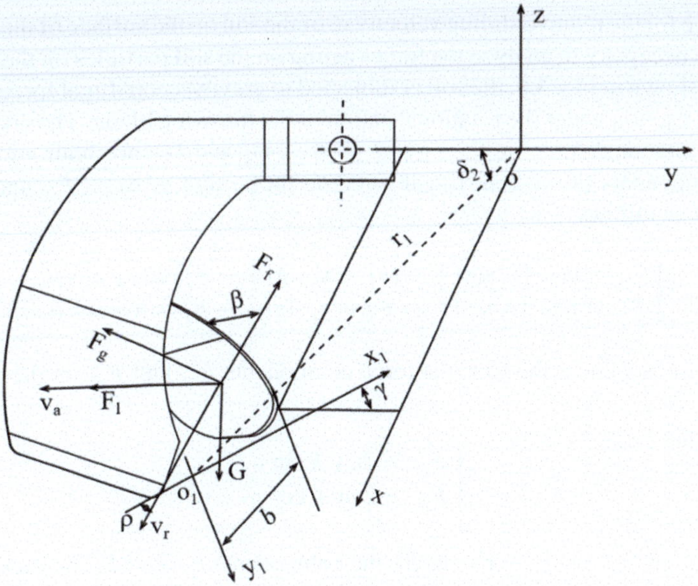

Fig. 3.9 Schematic diagram of the force of soil on the cast soil sheet

$$\begin{cases} v_x = v_r(\sin\gamma)(\sin\beta) + v_m \\ v_y = \frac{1}{2}v_r\sin(2a) - \frac{1}{2}v_r\sin(2\gamma)\cos\beta + v_a \\ v_z = v_r(\cos^2\gamma - v_r\sin^2\gamma\cos\beta + \omega r_1\cos(\omega t) \end{cases} \tag{3.12}$$

where: r_1—Point o_1 radius of rotation, m;

v_r—The relative speed of soil sliding on the surface of the cast sheet, m/s;

ω—Groove knife turning angular speed, rad/s;

v_m—The soil follows the speed of the trencher, m/s;

v_a—The speed at which the blade rotates to drive the soil, m/s.

v_x, v_y and, v_z are the components of absolute velocity v in the three-dimensional rectangular coordinate system of space. The absolute velocity v of soil on the soil throwing sheet is the vector sum of the implicated velocity v_a driven by the rotation of the soil throwing knife, the relative velocity v_r of soil sliding on the surface of the soil throwing sheet, and the velocity v_m of soil advancing with the ditching machine. Therefore, the vector sum can be obtained after solving it separately. The direction of the implicated velocity v_a is the negative direction of the Y-axis, and the calculation formula is as follows:

$$v_a = r_1\sin\left(\frac{\pi}{2} - \delta_2\right)\omega - x_1'\sin\left(\frac{\pi}{2} - \gamma\right)\omega \tag{3.13}$$

where: x_1—Instantaneous horizontal coordinates of soil particles on the thrown side;

δ_2—The Angle between r_1 and y, (°).

To solve for the relative sliding velocity v_r of the soil on the surface of the throwing blade, it is necessary to analyze the forces acting on the soil particles on the throwing blade. As shown in Fig. 3.8, the soil is subjected to gravity, centrifugal force, Coriolis force, and friction under the rotational action of the throwing blade. The components of the gravitational force on the x_1 and y_1 axes, G_x and G_y, are both mg sin γ cos β; the components of the centrifugal force on the x_1 and y_1 axes, F_{lx} and F_{ly}, are calculated as follows:

$$\begin{cases} F_{lx} = m\omega^2\left(r_1 \sin\left(\frac{\pi}{2} - \delta_2\right) - x_1' \sin\left(\frac{\pi}{2} - \beta\right)\right) \cos\gamma \cos\beta \\ F_{ly} = m\omega^2\left(r_1 \sin\left(\frac{\pi}{2} - \delta_2\right) - x_1' \sin\left(\frac{\pi}{2} - \beta\right)\right) \sin\gamma \cos\beta \end{cases} \quad (3.14)$$

Let $k = \cos\beta \cos\gamma$, the Coriolis force components F_{gx} and F_{gy} on the x_1 and y_1 axes are:

$$\begin{cases} F_{gx} = 2\omega v_r k \sin\rho \\ F_{gy} = 2\omega v_r k \cos\rho \end{cases} \quad (3.15)$$

Let $k_1 = \omega^2(r_1 \cos\delta_2 - x'_1 \cos\beta)$, the components F_{fx} and F_{fy} of friction force on the x_1 and y_1 axes are:

$$F_{fx} = mf \sin\rho(\sin\beta(g \sin\gamma + k) + 2v_r\omega(\cos\rho \sin\gamma + \sin\rho)\cos\beta) \quad (3.16)$$

where: ρ—the Angle between the relative velocity of soil and the x_1 axis, (°);

F—Friction factor between soil particles and the surface of the cast soil sheet.

In addition, according to the geometric relationship, the relative speed v_r direction satisfies:

$$\tan\rho = \frac{G_x + F_{lx} + F_{gx} + F_{fx}}{G_y + F_{ly} + F_{gy} + F_{fy}} = \frac{\sin\gamma}{\cos\gamma} \quad (3.17)$$

When the soil slides from the side cutting surface and tangential surface of the throwing blade to the y_1 axis of the throwing plate, the direction of the relative velocity v_r aligns with the x_1 axis. Hence, the angle ρ is 0°, and at this moment the soil is only subjected to gravity and centrifugal force. Under these conditions, Eq. (3.17) can be derived as follows:

$$g = -\omega \cos\beta\left(r_1 \cos\delta_2 + x_1' \cos\beta\right) \tan\gamma \quad (3.18)$$

Due to the end face height of the tangential surface of the throwing blade being 65 mm, and with the tangential edge and the side cutting edge curves remaining unchanged, the position of point o_1 can be determined. At this moment, the rotation radius r_1 of point o_1 is 230 mm, and the included angle δ_2 of the rotation radius at point o_1 is 60°. The throwing plate is made of manganese steel, and since the soil is mostly sandy loam, the friction coefficient f between the soil and the plate is 0.65. Given that the throwing plate is inclined along the bending line, similar to the tangential part, and maintains the same inclination angle, the angle β is consistent with the bending angle of the throwing blade, which is 60°. The machine's forward speed is taken as the commonly used speed of 0.5 m/s, and the rotational speed of the blade shaft is taken as the commonly used speed for ditching blades, which is 200 r/min, so ω is 20.94 rad/s. According to the agronomic requirements of throwing soil to a width of 300 mm at the edge of the ditch in the greenhouse, the throwing width S is set to 300 mm. Substitute the above parameters into Eqs. (3.12), (3.17), and (3.18) to solve for the unknowns γ, b, and v_r. By using the elimination method to remove v_r, we finally get $\gamma = 30°$ and $x'_1 = 120$ mm. At this time, the lateral soil throwing distance is 300 mm, so the designed width b of the throwing plate is equal to x'_1, which is 120 mm, ensuring a lateral soil throwing width of 300 mm.

3.2.5 *Simulation Test Model*

Using EDEM2020 software, we conducted a comparative analysis of the performance of both a single ditching blade and a combined cutting and throwing blade. The goal was to select the superior combined cutting and throwing blade. Subsequently, we performed a simulation analysis of the ditching performance of the chosen device, focusing on the effects of machine forward speed, blade shaft rotational speed, and ditch depth on ditching stability and power consumption. Based on the simulation results, we determined the optimal operating parameters to guide the design of machine parameters.

Simulation Model Parameter Design

When modeling soil using the discrete element method (DEM), it is crucial to accurately set the soil particle model. To ensure the accuracy of the simulation results, we measured the soil parameters within the greenhouse, specifically referencing the soil particle properties in a greenhouse located in the 6th Regiment of Aral City, Xinjiang. Additionally, we integrated research findings from other scholars on soil DEM parameters to establish a model for the interaction between the ditching blade and soil. The specific parameters are shown in Table 3.1.

Construction of Soil Particle Model

Using EDEM software, a soil model was established. Based on the agronomic requirements of the crops and the dimensions of the ditching and film-covering machine, the soil bin dimensions were set to 2000 mm × 600 mm × 400 mm (length × width × height). To simulate the characteristics of different soil particles in the greenhouse, three types of soil particle models were used: single-sphere, double-sphere, and triple-sphere, each with a radius of 1.5 mm. The contact model between soil particles was the Hertz-Mindlin with Bonding contact model, and the model was filled with soil particles under the influence of gravity. To enhance the bonding between particles, sedimentation was carried out after the initial filling, followed by

Table 3.1 Simulation model parameters

Item	Unit	Parameter value
Soil tank length × width × height	mm	2000 × 600 × 400
Forward speed	m/s	0.4–0.6
Speed of the trench cutter	r/min	180–220
Soil density	kg/m^3	1600
Poisson's ratio of soil		0.35
Soil shear modulus	pa	1×10^6
Iron density	kg/m^3	7860
Iron Poisson's ratio		0.4
Shear modulus of iron	pa	8×10^{10}
Soil-soil recovery coefficient		0.6
Soil-iron recovery coefficient		0.5
Soil-soil static friction factor		0.4
Soil-iron static friction factor		0.7
Soil-soil rolling friction factor	$N\ m^{-3}$	0.55
Soil-iron rolling friction factor	$N\ m^{-3}$	0.05
Stiffness coefficient of normal contact between particles	KPa	19,000
Tangential contact stiffness coefficient between particles	KPa	14,000
Critical normal stress between particles		50
Critical tangential stress between particles		30

continued filling until the model was completely filled. The specifics are shown in Figs. 3.10 and 3.11.

Construction of Ditching Blade Model

To improve simulation efficiency, the ditching blade model was simplified by removing structures that do not affect simulation results. Using SolidWorks 2023, two types of ditching blade models were established and saved in STEP format for import into EDEM 2020 software. In the preprocessing interface, the relative position of the ditching blade and the simulated soil was adjusted, and forward speed and rotational speed around the blade shaft were assigned to the ditching blade. The overall simulation model is shown in Fig. 3.12.

Simulation Experiment Design

This study aims to investigate the ditching performance of the ditching blade by simulating its movement within a soil bin. The simulation process is illustrated in Fig. 3.13. Based on design experience and keeping the field soil conditions constant, the ditching performance is primarily influenced by three factors: forward speed (A), blade shaft rotational speed (B), and ditch depth (C). Therefore, we designed a three-factor, three-level orthogonal experiment using the L9 (33) orthogonal table, with the

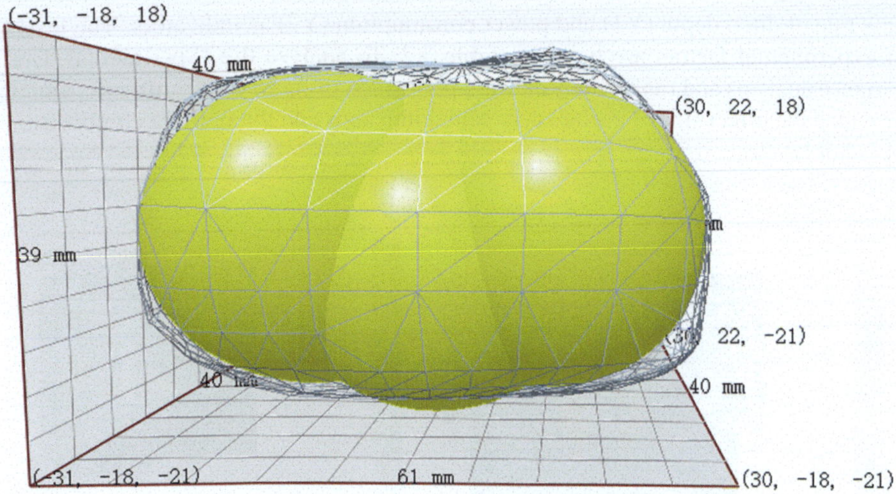

Fig. 3.10 The shape of the particle

Fig. 3.11 Simulated experimental soil tank

Fig. 3.12 Schematic diagram of the overall simulation model

pass rate of ditch depth (Y1) and power consumption (Y2) as indicators. The levels of experimental factors are listed in Table 3.2. To analyze the experimental data, variance analysis and range analysis were performed using SPSS27 software, which provided the effect size of each experimental indicator and the optimal combination of factor levels.

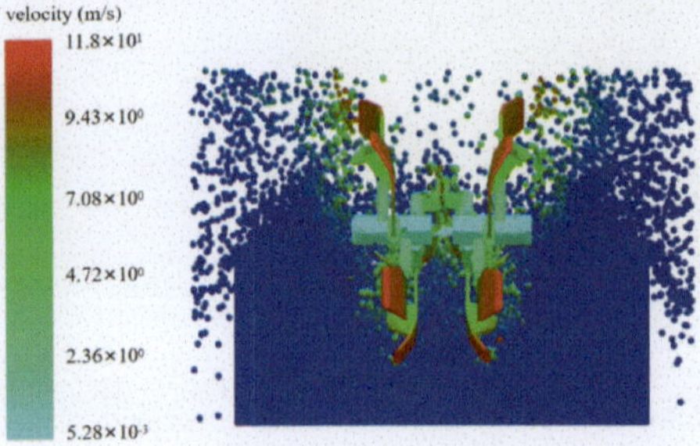

Forward speed v=0.4m/s, spindle speed n=180r/min, groove depth 30cm

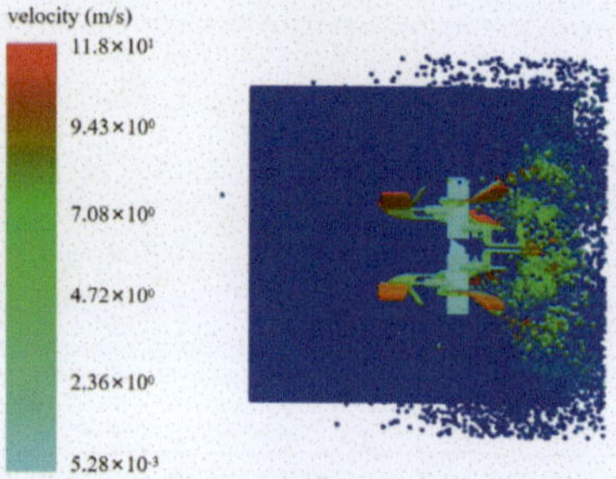

Forward speed v=0.4m/s, spindle speed n=200r/min, groove depth 30cm

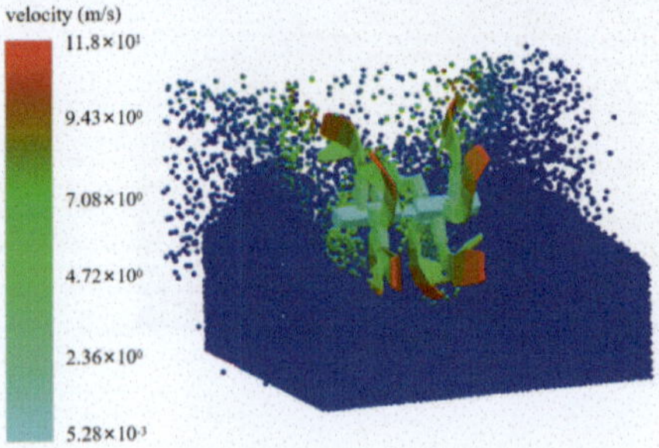

Forward speed v=0.5m/s, spindle speed n=180r/min, groove depth 30cm

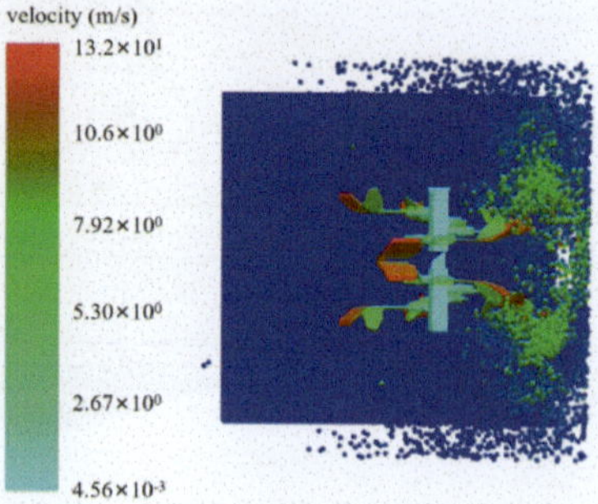

Forward speed v=0.5m/s, spindle speed n=180r/min, groove depth 30cm

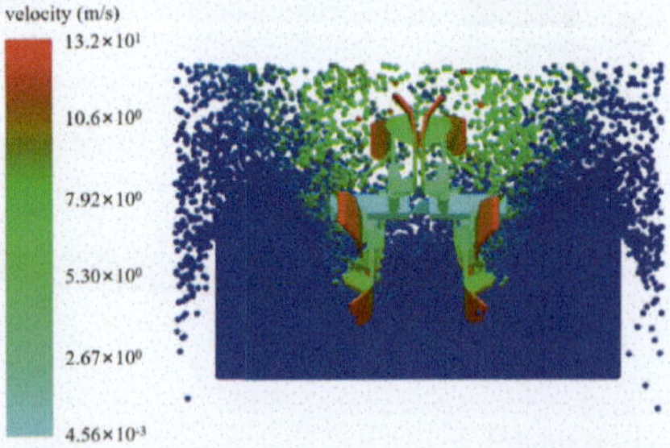

Forward speed v=0.5m/s, spindle speed n=200r/min, groove depth 30cm

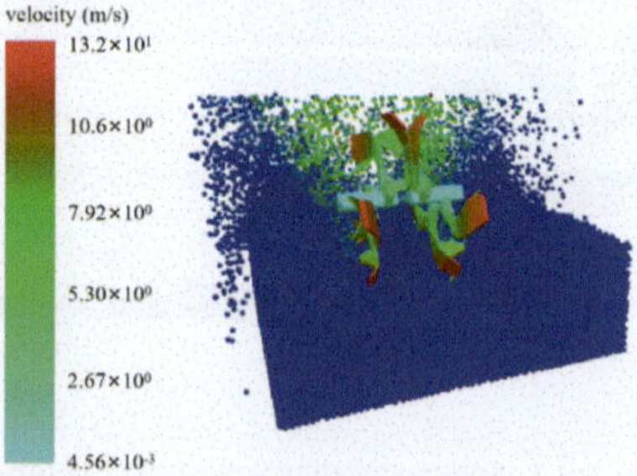

Forward speed v=0.5m/s, spindle speed n=220r/min, groove depth 30cm

Fig. 3.13 Simulation operation process of ditching tool under different working conditions

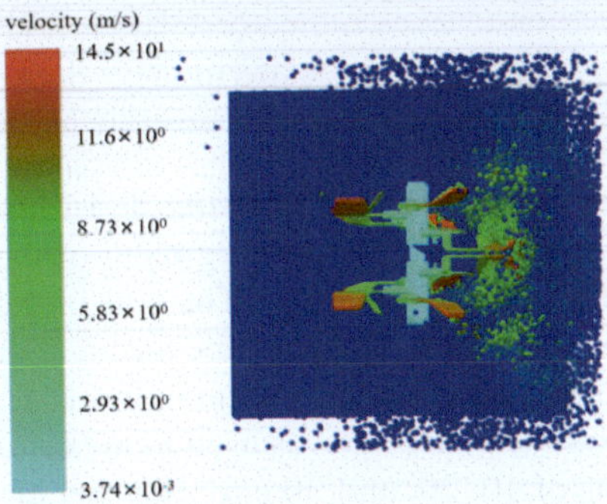

Forward speed v= 0.6m /s, spindle speed n=180r/min, groove depth 30cm

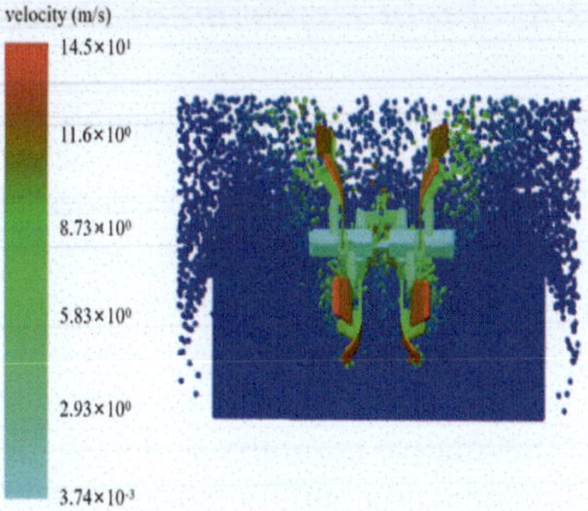

Forward speed v= 0.6m /s, spindle speed n=200r/min, groove depth 30cm

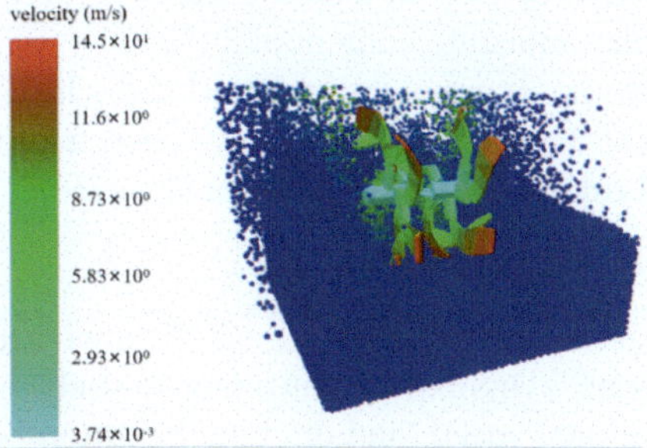

Forward speed v= 0.6m /s, spindle speed n=220r/min, groove depth 30cm

Fig. 3.13 (continued)

Table 3.2 Key parameters for the configuration of a single soil knife trench opener

Spindle speed (r/min)	Soil disturbance width (mm)	Return depth (mm)
180	435	47
200	426	46
220	431	49

The ditching process is crucial for the formation of the final trench characteristics. To study the impact of different experimental factors on trench stability, the cross-section of the trench was fitted during the simulation process and presented as marked grid coordinate points. Additionally, the trench contour curves for different ditching blades under various experimental factors were plotted. We analyzed the backfill depth and soil disturbance and calculated the ditching stability and trench width consistency coefficient for each group of simulations involving the combined cutting and throwing ditching blade.

3.2.6 *Field Experiment Design*

Experimental Conditions and Materials

In September 2023, a field experiment was conducted with a ditching and film-covering integrated machine in the experimental field of the Sixth Regiment of the First Division of the Xinjiang Production and Construction Corps. The soil in the experimental field was sandy, and consistent soil conditions were ensured throughout the experiment. During the experiment, the temperature was approximately 21 °C, and the relative humidity was 41%. The five-point test method was used to measure soil compaction and moisture content at depths of 10, 20, and 30 cm. The specific data are as follows: the average soil compaction at a depth of 10 cm was 142.3 with a moisture content of 11.4%; at a depth of 20 cm, the average soil compaction was 163.4 with a moisture content of 11.3%; and at a depth of 30 cm, the average soil compaction was 187.6 with a moisture content of 11%.

The equipment used in the experiment included a ditching and film-covering integrated machine, a TYD-2 digital soil hardness tester, a soil moisture meter, a steel tape measure, a steel ruler, and an electronic balance (Fig. 3.14).

Ditching Power Consumption Test

The test methods were based on the NY/T740—2003 standard, "Quality of Field Ditching Machine Operations." This experiment aimed to study the effects of forward speed, ditching blade rotational speed, and ditching depth on power consumption, using power consumption as the primary indicator. In the field tests, the operational speed of the machine and the rotational speed of the blade were determined based

Fig. 3.14 Field test of ditching and laminating machine

on the values listed in the table, influenced by the gear position and throttle settings. Adjustments to the machine's forward speed and blade shaft rotational speed were made through gear and throttle settings, while the ditching depth was controlled by adjusting the relative height of the depth-limit wheel's threaded rod and the fixed sleeve. Power consumption was measured using a torque sensor to capture the instantaneous torque of the blade, from which instantaneous power was calculated. After the test, the data processing terminal recorded the average power over the entire operation path. By analyzing the data from the table of experimental factors, we examined the interaction effects of each factor on ditching power consumption.

3.3 Results and Discussion

3.3.1 Simulation Experiment Results and Analysis

Simulation of Trench Cross-Section Results and Analysis

This experiment utilized the discrete element method (DEM) for simulations. The ditching process is crucial for the formation of the final trench characteristics. To study the impact of various experimental factors on trench stability, the simulated trench cross-sections were fitted using marked grid coordinate points. Subsequently, the trench contour curves were plotted under the influence of different experimental factors, as shown in Fig. 3.15.

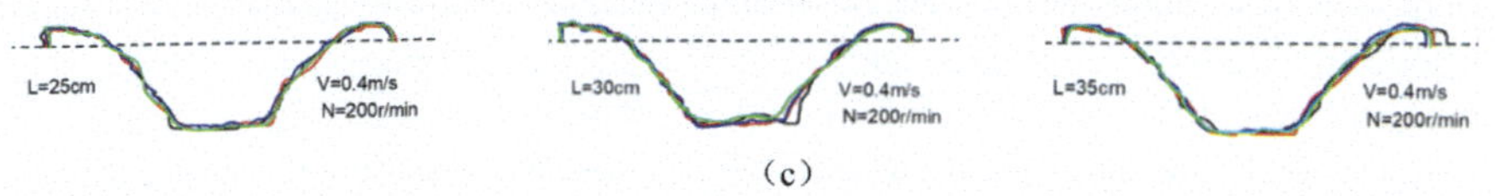

Fig. 3.15 Influence of experimental factors on furrow type and furrow ridge. **a** The effect of advancing speed on trench pattern and ridge; **b** The influence of spindle speed on groove type and groove ridge; **c** Influence of trench depth on trench type; Black line—is the outline of the trench ridge of the combined throw-cut trench cutter, with the cross section at the center point. The combined blue wireline and throw-cut trench cutter is the contour of the trench ridge at 0.5 m away from the center point. Green line—outline of a single trench cutter trench ridge, cross section at the center point. Red line—a single cutter is the outline of the furrow ridge at 0.5 m from the center point

As the forward speed increases, the groove shape becomes unstable due to the increased inertial force exerted on the soil, leading to uneven soil movement and accumulation. The design of the single-cutting blade relies solely on a single cutting method, lacking mechanisms for multi-point support and pressure dispersion. Under these conditions, it is more prone to causing extensive soil disturbance, resulting in significant soil particle accumulation on both sides of the groove and irregular groove shapes. In contrast, the throw-cut combined trenching blade operates with multiple blades working together to distribute forces more evenly. Its unique design minimizes soil throw and fallback, resulting in a more stable groove shape.

As the rotational speed of the blade shaft increases, the groove shape initially stabilizes and then becomes unstable. At lower initial speeds, the cutting energy is insufficient, preventing effective soil separation and causing local accumulation. When the speed reaches a critical value, the cutting energy becomes optimal, allowing uniform soil cutting and separation, forming a relatively stable groove shape. However, further increasing the speed results in excessive cutting force, causing the soil to be rapidly thrown away from the cutting point, leading to excessive disturbance and making the groove shape unstable again. In comparison, the throw-cut combined trenching blade can segment the soil at multiple points, reducing the force at each point and extending the stable speed range, thereby maintaining superior groove shape stability over the single-cutting blade.

As the groove depth increases, the forces exerted by the cutting blade on the soil increase, leading to more significant soil disturbance and displacement, thus making the groove shape increasingly unstable. Unlike shallow trenching operations, increased depth results in a thicker soil layer, enhancing both vertical and horizontal soil displacement. At this point, both the throw-cut combined trenching blade and the single-cutting blade exhibit a trend of decreasing stability with increasing groove depth. This indicates that at greater depths, the impact of blade design on soil disturbance tends to converge. Nevertheless, the throw-cut combined trenching blade still demonstrates advantages under most conditions.

Analysis of Backfill Depth and Soil Disturbance

Based on data from Tables 3.2 and 3.3, it can be observed that, at the same forward speed, as the rotational speed of the blade shaft increases, the soil disturbance caused by both the single-cutting blade and the throw-cut combined trenching blade also increases, indicating that the rotational speed of the blade shaft has a significant impact on soil disturbance. At high rotational speeds, soil experiences greater disturbance and movement, affecting the trenching quality. The maximum soil disturbance width for the single-cutting blade is 435 mm, and the maximum backfill depth is 49 mm. In comparison, the throw-cut combined trenching blade exhibits a maximum soil disturbance width of 413 mm and a maximum backfill depth of 41 mm. The data indicate that the single-cutting blade causes greater soil disturbance during operation, leading to more extensive soil disruption. In contrast, the throw-cut combined trenching blade demonstrates smaller disturbance width and depth, suggesting reduced soil disturbance and better groove shape maintenance during trenching. The throw-cut combined trenching blade, through its design characteristics, ensures that soil is evenly thrown out of the trench, preventing the soil from falling back into the groove. This is crucial for maintaining the width and shape of the groove bottom. Conversely, the single-cutting blade, due to its design limitations, tends to throw soil back into the trench, causing backfill and affecting the groove shape.

The throw-cut combined trenching blade shows superior performance in controlling soil disturbance compared to the single-cutting blade, achieving better trenching results and soil disturbance control. Its design effectively reduces soil disturbance width and backfill depth. With its smaller soil disturbance width and depth, and better soil throwing effect, it helps maintain a suitable groove bottom width, achieving high-quality trenching results and providing a better planting environment. This is critical for meeting the requirements of the embedded substrate cultivation model. Therefore, the throw-cut combined trenching blade is the more ideal choice.

The data were organized and plotted using Origin2022 software to study the effects of two types of trenching blades on groove shape parameters. Figure 3.16 shows the comparison of groove shape parameters formed by the two trenching blades under different operating conditions.

The trenching depth stability and trench width consistency coefficient for each set of simulated experiments using the throw-cut combined trenching blade were calculated, as shown in Fig. 3.13. The calculation method for trenching depth stability

Table 3.3 Key parameters of the combined throw-cut trench opener configuration

Knife shaft rotational speed (r/min)	Soil disturbance width (mm)	Backfill depth (mm)
180	404	39
200	396	41
220	413	37

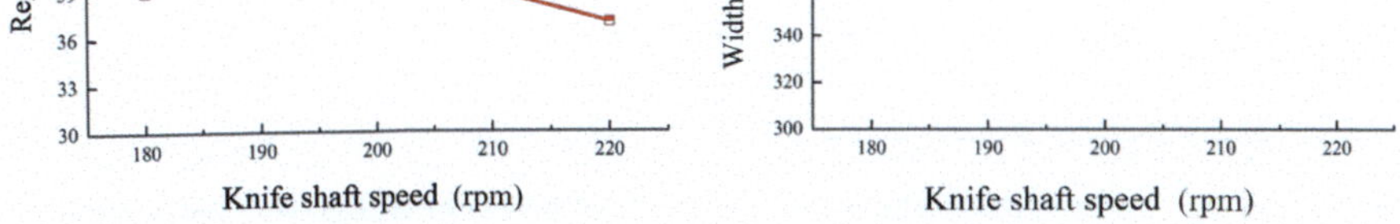

(a) soil return depth comparison (b) Soil disturbance width correlation

Fig. 3.16 Analysis of trenching stability and trench width consistency coefficient is as follows:

$$L_1 = \frac{L_0}{L} \times 100\% \tag{3.19}$$

where: L_1—Stability rate of ditching depth (%);

L_0—Actual trench depth (cm);

L—Theoretical trenching depth (cm).

As shown in Fig. 3.17a, at lower forward speeds, the tool can fully contact the soil and perform effective cutting, resulting in minimal and evenly distributed soil disturbance. However, when the speed increases from 0.4 to 0.55 m/s, the soil stress increases, causing the groove shape to become irregular due to the disturbance, resulting in a significant decrease in groove stability. When the forward speed increases to 0.6 m/s, the inertial force applied by the tool to the soil increases, resulting in accelerated movement of soil particles, exacerbating soil rolling and irregular deposition. The contact time between the tool and the soil is shortened, and the tool cannot fully overcome the soil resistance, reducing the groove stability. Nevertheless, when the speed reaches 0.6 m/s, the groove width consistency remains at 88.5%, as the soil has basically reached a balanced state, and even if there is disturbance, the structure remains relatively stable.

Figure 3.17b shows that at lower blade shaft rotational speeds, the cutting energy provided by the tool is insufficient to overcome the soil cohesion, causing uneven soil stress and resulting in an unstable groove structure. As the blade shaft rotational speed increases, the cutting efficiency improves, and the soil can be quickly and evenly separated, leading to a more stable groove structure. Groove stability gradually increases, peaking at 200 r/min, with trenching depth and groove width consistency exceeding 85%. However, when the rotational speed continues to increase to 210 r/min or above, the impact and frictional forces between the tool and the soil significantly increase, easily causing secondary soil disturbance and displacement.

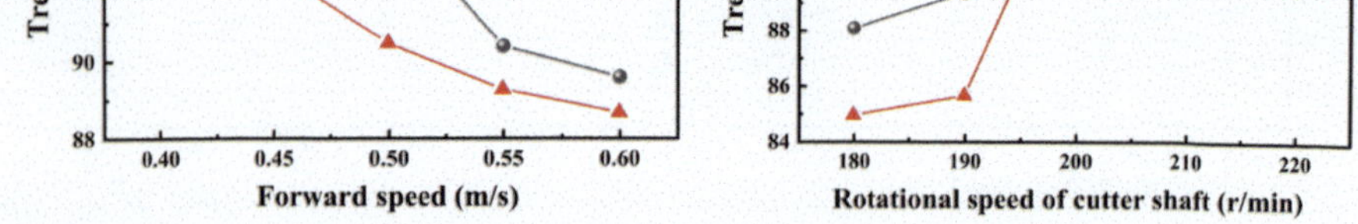

Fig. 3.17 (a) As the forward speed increases. (b) As the cutter shaft speed increases

Secondary rolling of the soil also decreases the consistency of groove depth and width, thereby affecting the overall groove stability.

Results and Analysis of the Orthogonal Simulation Experiment for the Trenching Blade

The power consumption of the trenching device primarily consists of two parts: one is the torque power required for blade shaft rotation, and the other is the tractive power required for the forward movement of the trenching device. The sum of these two components represents the total power consumption of the trenching device. This study used the discrete element method (DEM) to numerically calculate and analyze the torque of the trenching blade under different motion parameters.

The EDEM software was used for the discrete element simulation of the torque of the trenching blade to study its power consumption characteristics. The built-in post-processing module Analyst Tree in the software was used to export the torque and force data during the simulation of the trenching blade, with the data collection interval set to 0.1 s to ensure data accuracy and continuity. Based on the set rotational speed constants and the torque and force obtained from the simulation experiments, the power consumption of the trenching blade was calculated using formulas.

In the simulation tests, the motion trajectory of the trenching blade in the soil trough is shown in Fig. 3.18. The trenching blade rotates around its axis and moves uniformly along the y-axis direction. To eliminate the transient effects before and after the trenching blade enters and leaves the soil trough, only the time interval from the tool's full contact with the soil to its departure from the soil was analyzed to reflect the actual working state of the trenching blade. Figure 3.18 shows the trenching effect after the operation was completed.

The torque power consumption and traction power consumption of the tool, calculate the total power consumption of the trenching process:

$$P_1 = \frac{T \times n}{9550} \tag{3.20}$$

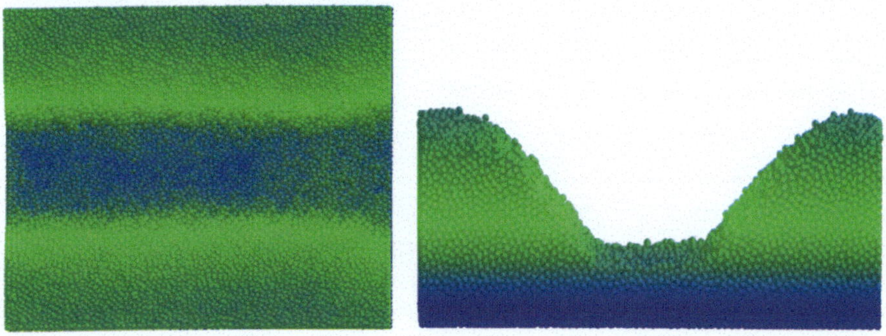

Side throw effect   Floating soil at the bottom of the ditch

Fig. 3.18 Ditching effect

where: T—Torque acting on the tool (N);

n—Tool speed.

The calculation formula of traction power consumption is as follows:

$$P_2 = F \times v \tag{3.21}$$

where: F—Resistance applied to the cutter (N);

v—Furrow cutter advance speed (m/s).

In other words:

$$P = P_1 + P_2 \tag{3.22}$$

SPSS data processing software was used to analyze the experimental results, as shown in Tables 3.4 and 3.5.

The data analysis results from Tables 3.4 and 3.5 show that among the three test factors affecting the trenching qualification rate, the traction speed, trenching depth,

Table 3.4 Analysis of orthogonal experiment results

Test number	Forward speed A (m/s)	Spindle speed B (r/min)	Trenching depth C (cm)	Pass rate of ditch depth Y1 (%)	Power dissipation Y2 (W)
1	1 (0.4)	1 (180)	1 (25)	97.6	655
2	1	2 (200)	2 (30)	99.2	515
3	1	3 (220)	3 (35)	97.8	867
4	2 (0.5)	1	2	98.2	972
5	2	2	3	97.7	1008
6	2	3	1	98.4	892
7	3 (0.6)	1	3	96.3	1513
8	3	2	1	97.3	1032
9	3	3	2	98.1	1347
K1	294.89	289.35	293.30		
K2	290.38	292.62	293.45		
K3	290.71	294.01	289.23		
k1	98.30	96.45	97.77		
k2	96.79	97.54	97.82		
k3	96.90	98.00	96.41		
Range R	1.50	1.55	1.41		
Optimum level	A1	B3	C2		
Primary and secondary factors		B > A > C			

Table 3.5 Orthogonal experiment results

Index	Difference source	Sum of squares	Degree of freedom	Mean square	f-value	Significance
Pass rate of ditch depth (Y1)	A	2.069	2	1.035	129.313	0.007
	B	1.029	2	0.514	64.306	0.015
	C	2.309	2	1.154	144.306	0.008
	Error	0.016	2	0.008		
	Sum total	5.423	2			
Power dissipation (Y2)	A	573,889.8	2	286,944.9	168.103	0.06
	B	72,673.88	2	36,336.94	24.414	0.045
	C	114,297.9	2	57,148.95	33.679	0.029
	Error	3393.736	2	1696.868		
	Sum total	764,255.3	8			

Table 3.6 Test factor levels

ID	Factor		
	Forward speed (m/s)	Spindle speed (r/min)	Trench depth (cm)
1	0.4	180	25
2	0.5	200	30
3	0.6	220	35

and blade shaft rotational speed have the most significant impact on the experimental results, with the blade shaft rotational speed having the most significant effect. The importance of the test factors is ranked as follows: blade shaft rotational speed, forward speed, and trenching depth.

3.3.2 *Field Test Results and Analysis*

The levels of the test factors are shown in Table 3.6, and the test schemes and results are shown in Table 3.7.

3.3.3 *Response Surface Analysis*

Using Design-Expert 13 to fit the experimental data, the regression equation for the effect of forward speed (v), blade shaft rotational speed (n), and trenching depth (h) on the trenching power consumption (p) is shown in Eq. (3.23):

Table 3.7 Test scheme and results

Test number	Forward speed (m/s)	Spindle speed (r/min)	Trench depth (cm)	Power consumption (kW)	Ditch depth stability (%)
1	0.6	220	30	0.751	78.6
2	0.5	200	30	0.668	84.3
3	0.4	200	35	0.751	80.2
4	0.5	180	25	0.614	84.3
5	0.5	180	35	0.657	84.6
6	0.5	220	35	0.692	81.4
7	0.6	180	30	0.716	82.6
8	0.4	180	30	0.716	76.4
9	0.5	200	30	0.677	85.4
10	0.5	200	30	0.653	83.6
11	0.6	200	35	0.742	82.4
12	0.4	200	25	0.647	80.4
13	0.4	220	30	0.726	78.3
14	0.5	200	30	0.662	86.3
15	0.5	200	30	0.674	86.7
16	0.5	220	25	0.638	84.1
17	0.6	200	25	0.712	83.6

$$P = 0.6668 + 0.0101A + 0.0129B + 0.0288C - 1.48AB - 0.25AC - 0.75BC - 4.12A^2 - 2.17B^2 - 0.0155C^2 \quad (3.23)$$

The regression equation for the effect of forward speed (v), blade shaft rotational speed (n), and trenching depth (h) on the trench depth stability (s) is shown in Eq. (3.24):

$$S = 85.26 + 1.49A - 0.6875B - 0.475C - 1.48AB - 0.25AC - 0.75BC - 4.12A^2 - 2.17B^2 + 0.5075C^2 \quad (3.24)$$

where V represents trenching power consumption, S represents trench depth stability, A represents operation speed, B represents blade shaft rotational speed, and C represents trenching depth.

To clearly demonstrate the effect of different experimental factors on the trenching power consumption and trench depth stability of the machinery, we fitted the quadratic regression model equations for trenching power consumption and trench depth stability (Eqs. 3.5–3.7 and 3.5–3.8). The resulting response surface optimization models are shown in Figs. 3.19 and 3.20.

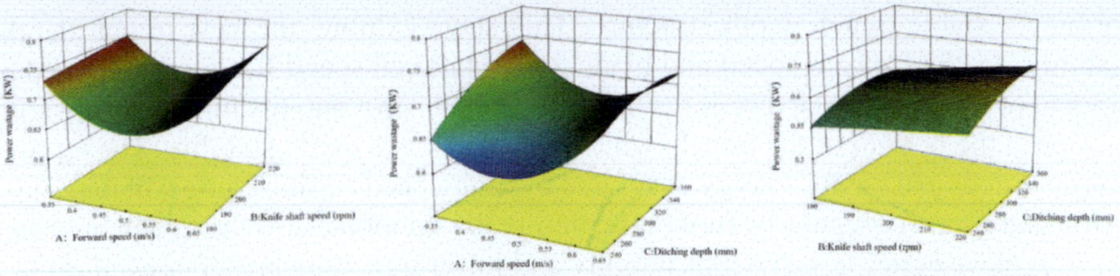

a. A and B interact with response surfaces and contours

b.A and C interact with response surfaces and contours

c.B and C interact with response surfaces and contours

Fig. 3.19 Power response surface and contour diagram of each factor

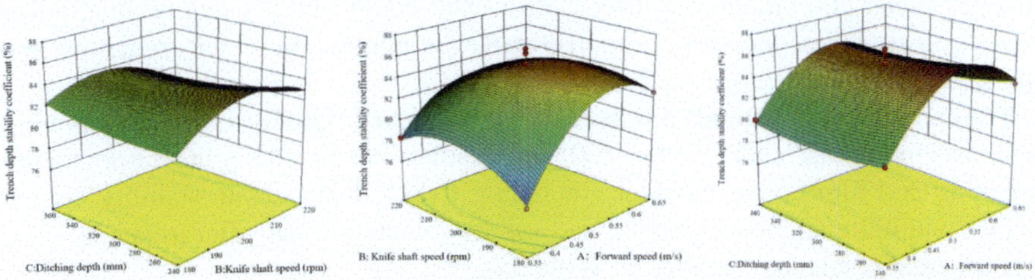

a. A and B interact with response surfaces and contours

b. A and C interact with response surfaces and contours

c. B and C interact with response surfaces and contours

Fig. 3.20 Response surface and contour diagram of each factor to the stability of ditch depth

According to Fig. 3.19, when trenching depth remains constant, an increase in the forward speed of the machinery and the rotational speed of the blade disk leads to higher power consumption. Among these factors, the blade shaft rotational speed has the most significant impact on power consumption. An increase in forward speed results in more soil particles being cut and transported per unit time, requiring the trenching blade to handle more soil, thereby increasing cutting resistance and friction, which in turn raises power consumption. Higher blade shaft rotational speeds increase the frequency of contact between the trenching blade and soil particles, leading to more soil particles being cut and thrown out. This also enhances the adhesion and friction between the blade surface and soil particles. These factors collectively increase energy consumption during the cutting and transporting process, further elevating power consumption. Through contour plots and response surface diagrams, as well as interaction analysis between factors using the range analysis function of the software, the primary and secondary order of the factors affecting power consumption is determined as: blade shaft rotational speed > operation speed > trenching depth.

According to Fig. 3.20, at lower forward speeds and blade disk rotational speeds, the cutting action of the tool on the soil is weaker, making it difficult for soil particles to be thrown out of the trench. This results in less soil residue at the bottom of the trench and higher trench depth stability. When the speed and rotational speed reach moderate levels, the cutting and throwing actions of the tool on the soil reach a balanced state. The amount of soil being cut and thrown out is relatively balanced, forming a stable trench depth structure. This stage exhibits the best trench depth stability. As the forward speed and blade disk rotational speed further increase, the amount of soil being cut exceeds the amount being thrown out, resulting in more soil residue at the bottom of the trench, which reduces trench depth stability. The strong inertial forces and kinetic energy generated by high-speed cutting cause soil particles to roll and deposit irregularly, further disrupting the smoothness of the trench bottom.

To achieve better trench depth stability and lower trenching power consumption, the optimal working parameters were selected based on the agronomic requirements of embedded substrate cultivation: an operating speed of 0.5 m/s, blade shaft rotational speed of 200 r/min, and trenching depth of 30 cm. Under these parameters, the trenching power consumption is 0.668 kW, and the trench depth stability is 86.7%. This parameter combination ensures good trench depth stability and relatively low power consumption, meeting the performance requirements of the trenching and film-laying machine.

The experimental methods followed the guidelines of "Quality of Field Trenching Machine Operations" (NY/T740—2003). This study calculated key indicators for the trenching and film-laying machine, including trench depth, depth stability, trench surface width, trench bottom width, and soil fragmentation rate after trenching. The experimental results are detailed in Table 3.8.

The method for measuring trench depth is as follows: Before testing, clear the soil at the bottom and around the trench. Select two operational paths, each no less than 50 m long. Along the operational direction, select five evenly spaced points for measurement, making a total of ten points. Place a ruler at the intersection of the original ground surface and the two trench walls. The trench depth at each measurement point is the vertical distance from the trench bottom center to the ruler.

Average trench depth:

$$h = \frac{\sum_{i=1}^{n} h_i}{N} \tag{3.25}$$

where: h—Average trenching depth, cm;

h_i—Trench depth value at the i-th measurement point, cm;

Table 3.8 Results of field experiment in ditching

Operational performance index	Unit	Standard	Experimental measurement
Trenching depth	cm	About 30 (agronomic requirements)	31
Ditching depth stability	%	≥80.0	87.1
Channel width	cm	40	41.3
Width of trench	cm	30	30.4
Rate of broken soil after ditching	%	≥55	92.4

N—The number of measurement points selected in the work area;
$N = n = 10$.
The calculation formula for the stability of ditching depth is as follows:

$$S = \sqrt{\frac{\sum_{i=1}^{n} (h_i - h)^2}{N - 1}} \tag{3.26}$$

$$V = \frac{S}{h} \times 100\% \tag{3.27}$$

$$U = 1 - V \tag{3.28}$$

where:
S—Standard deviation of trench depth, cm;
V—Coefficient of variation of trench depth, %;
U—Stability coefficient of trench depth, %;
$N = n = 10$.

Through field experiments, measurements and calculations of trench depth stability and soil fragmentation rate were conducted. Data on trench depth, trench shape, trench surface width, and trench bottom width were collected, as shown in Table 3.8. By comparing the experimental results with local agronomic requirements and relevant standards, the following conclusions were drawn. The average trench depth in the field experiments was 31 cm, which met the standards. The depth fluctuation range was small, indicating good stability and reliability of the trenching knife. The average trench depth stability was 87.1%, higher than the standard of 80.0%. The small depth variation is beneficial for subsequent farming and cultivation operations. The average trench surface width was 41.3 cm, and the trench bottom width was 30.4 cm. This helps enhance soil drainage and aeration, improving the quality of the crop growth environment. The average soil fragmentation rate after trenching was 92.4%. A high fragmentation rate indicates that the trenching machine can fully break the soil during operation, enhancing soil looseness. This is very beneficial for root growth and water and nutrient penetration, and it also reduces the difficulty and labor intensity of subsequent land preparation. Additionally, a comparison of simulation test results with field test results showed a high degree of consistency under the same parameter conditions, validating the accuracy of the simulation experiments. In conclusion, the trenching knife motion analysis and simulation methods used in this study are reliable and reasonable.

The Northwest region is predominantly composed of deserts, gobi, and saline-alkali lands, where crops are typically cultivated in solar greenhouses. However, traditional ditching and mulching machines are large in size and perform poorly in sandy soils, leading to inferior trenching and soil-throwing, which ultimately reduces crop yield and affects economic returns. Therefore, it is crucial to design a cutting and throwing combined ditching blade, which can enhance the mechanized planting level of crops under the solar greenhouse sandy substrate cultivation model. Initially, a spiral staggered arrangement of the cutting and throwing combined ditching blade was designed. We analyzed the interaction between soil and the soil-throwing blade, established the dynamic equations of soil particles on the blade surface, determined key parameters, and developed an EDEM simulation model. Simulations were conducted to compare single ditching blades with combined blades, resulting in the selection of the optimal ditching blade. Subsequently, a three-factor, three-level orthogonal experiment was performed to identify the most significant factors affecting ditching performance. Finally, a prototype was manufactured, and field trials were conducted.

1. The overall design of the integrated ditching and mulching machine was completed, and the key component—the ditching blade—was optimized. Key dimensional parameters were determined by analyzing the soil-throwing process and the forces acting on the soil-throwing blade. The optimization results are as follows: the height of the ditching blade's end surface in the tangential plane was increased to 65 mm, the inclination angle of the soil-throwing blade was set to 30°, the width was set to 12 cm, and the transverse soil-throwing width reached 30 cm.
2. Using EDEM software for simulation modeling, the trench cross-section was curve-fitted and displayed in the form of marked grid coordinate points. By analyzing the backfill depth and soil disturbance of two types of ditching blades, the optimal ditching blade was determined. Further calculations analyzed the trenching stability and trench width consistency coefficient of the combined cutting and throwing ditching blade in various simulation experiments. At the maximum forward speed of 0.6 m/s, the trench width consistency was 88.5%, meeting trenching stability standards. The maximum soil disturbance width was 413 mm, and the maximum backfill depth was 41 mm. Compared to the single ditching blade, soil disturbance was reduced by 5.3%, and backfill depth was reduced by 19.5%. These results indicate that the combined cutting and throwing ditching blade performs better in resisting soil disturbance. Finally, through orthogonal experimental design, the significant factors affecting ditching performance were identified. The optimal parameter combination was obtained: operation speed of 0.5 m/s, blade shaft speed of 200 r/min, and trenching depth of 30 cm. Under these parameters, trenching power consumption was 0.668 kW, and trench depth stability was 86.7%.
3. Field test results showed that the average trenching depth of the equipment was 31 cm, meeting agricultural production standards. The stability of trench depth averaged 87.1%, the trench surface width averaged 41.3 cm, and the trench bottom width was 30.4 cm. The average soil fragmentation rate after trenching was 92.4%, with all key indicators meeting national standards. The equipment demonstrated good trenching effect and stability, solving the problem of poor trenching quality and unsatisfactory soil-throwing performance in the sandy substrate cultivation model in the Northwest region, thereby improving agricultural production efficiency.

The integrated ditching and mulching machine designed in this study differs significantly from traditional machines in its smaller size, making it suitable for use in solar greenhouses in non-arable areas. Additionally, the optimized ditching blade is better suited for trenching operations in sandy soils. Therefore, this study provides an important reference for mechanized operations in the solar greenhouse sandy substrate cultivation model. Given that the Northwest region predominantly consists of sandy soils, the ditching blade needs to operate at high frequency. Further research on the blade material is necessary to enhance its wear resistance, extend its service life, and improve performance stability, which will be a key focus for future work.

Chapter 4
Study on Intelligent Identification and Transplanting Technology of Plug Seedlings in Desert and Gobi Areas

Jianfei Xing, Long Wang, Zhicheng Liu, Xuyang Ran, and Shan Jiang

4.1 Introduction

Under the background of global warming and desertification, agricultural production faces significant challenges, particularly in desert and Gobi regions where harsh natural conditions, poor soil, and scarce water resources make traditional farming methods difficult to scale. Therefore, leveraging modern technology to enhance agricultural productivity and reduce resource waste has become an urgent issue. In this context, the integration of intelligent plug seedling identification and transplanting technology has gradually emerged as a hotspot in agricultural technology development, bringing new hope for agricultural production in desert and Gobi areas.

J. Xing · L. Wang (✉) · Z. Liu · X. Ran
College of Mechanical and Electrical Engineering, Tarim University, Alar 843300, China
e-mail: 120140002@taru.edu.cn

Modern Agricultural Engineering Key Laboratory at Universities of Education Department of Xinjiang Uygur Autonomous Region, Alar 843300, China

Xinjiang Production and Construction Corps (XPCC) Key Laboratory of Utilization and Equipment of Special Agricultural and Forestry Products in Southern Xinjiang, Alar 843300, China

J. Xing
e-mail: 120200012@taru.edu.cn

Z. Liu
e-mail: 10757222241@stumail.taru.edu.cn

S. Jiang
Key Laboratory of Smart Agriculture System Integration, Ministry of Education, Beijing 100083, China

Key Laboratory of Agricultural Information Acquisition Technology, Ministry of Agriculture and Rural Affairs, Beijing 100083, China

C. Hu et al. (eds.), *New Technologies Applied in Gobi Desert Facility Agriculture*, Smart Agriculture 11, https://doi.org/10.1007/978-981-96-1031-0_4

As an essential part of agricultural automation, intelligent recognition technology has made remarkable progress in recent years. By utilizing cutting-edge technologies such as computer vision, machine learning, and deep learning, intelligent identification systems can accurately recognize the growth status, variety, and health of plug seedlings, as well as determine the optimal timing for transplanting. This technology not only reduces human error and improves production efficiency but also provides real-time monitoring and feedback, offering scientific support for agricultural production and helping farmers better manage the planting process. In the unique ecological environment of desert and Gobi regions, where crop growth conditions are more demanding, intelligent identification technology is particularly crucial. It enables precise monitoring and control, ensuring a higher success rate for transplanting.

Transplanting technology is a critical component of agricultural production, especially in desert and Gobi areas, where crop survival rates are of utmost importance. Traditional transplanting methods rely heavily on manual operations, which, although flexible, are inefficient and prone to errors. With the advancement of agricultural automation, modern transplanting technology has increasingly become intelligent and automated, employing machinery to perform large-scale, efficient transplanting operations, significantly improving transplanting efficiency and crop survival rates. This technology not only saves substantial labor resources but also provides favorable conditions for healthy crop growth by precisely controlling key parameters such as transplanting depth and distance.

The combination of intelligent recognition and transplanting technology has led to the development of an integrated solution for intelligent plug seedling identification and transplanting, representing a major breakthrough in agricultural technology. The core of this technology lies in real-time monitoring of plug seedlings' growth and transplanting needs through intelligent recognition systems, with data transmitted to automated transplanting equipment for precise and rapid operations. Compared to traditional methods, this technology has made qualitative leaps in both efficiency and precision, and can also handle individual seedlings based on specific conditions, further increasing crop survival rates.

In desert and Gobi areas, the application of intelligent plug seedling identification and transplanting technology holds broad prospects. Firstly, it can effectively cope with harsh natural environments by reducing seedling damage during transplanting through intelligent monitoring and precise operations, thereby improving survival rates. Secondly, the high degree of automation and intelligence significantly reduces labor costs, meeting the demands of large-scale agricultural production. Finally, by integrating big data and artificial intelligence technologies, future intelligent plug seedling identification and transplanting systems could be further optimized, enhancing their self-learning and adaptive capabilities, and driving agricultural production toward greater intelligence.

In conclusion, the research on intelligent plug seedling identification and transplanting technology is not only an innovation in modern agricultural science and technology but also a powerful measure to address global climate change and desertification challenges. The successful application of this technology will provide strong technical support for agricultural production in desert and Gobi regions, promoting agricultural development towards greater efficiency, intelligence, and sustainability. In future research, how to further enhance the integration of intelligent identification and transplanting technologies, optimize system performance, and promote widespread application will be key issues for researchers.

4.2 Research Status of Intelligent Identification and Transplanting Technology for Plug Seedlings in China and Abroad

4.2.1 Research Status of Intelligent Identification of Plug Seedlings Abroad

With the rapid development of computer technology, machine vision has been increasingly applied in the transplanting process of plug seedlings. Through machine vision, seedlings can be identified, and subsequent transplanting operations can be executed based on the identification results, effectively reducing the rate of missing transplants and improving overall transplanting outcomes.

As early as 1994, Tai et al. from Ohio State University in the United States developed an automatic transplanter assisted by machine vision. This system utilized image processing technology to identify and locate empty cells in the plug tray and tested 15 different types of seedlings. The results showed that the algorithm could effectively detect seedlings with uneven growth heights in the trays, achieving a 95% success rate in identifying empty cells. This research laid the foundation for the future application of machine vision technology in agricultural automation.

In 2001, RYU et al. from Seoul National University in South Korea developed a transplanting robot based on a machine vision system. The robot mainly consisted of a conveyor belt, a grabbing manipulator, a planting mechanism, and a machine vision system. Due to the significant color difference between the seedling leaves and the background of the tray, the machine vision system adopted a color-based image segmentation method to capture the leaf area. The system then compared the leaf area with preset thresholds to locate the empty cells, greatly enhancing transplanting efficiency. In the same year, the Ferrari company in Italy developed a transplanter equipped with a machine vision system that used a laser-based photoelectric sensor to detect whether there was a seedling in the transplanter's seedling device. If a seedling was detected, the next step would proceed; if not, the device would automatically rotate and adjust.

Additionally, the Dutch company TTA developed the Flex series and Midi series of fully automated transplanters with machine vision capabilities. The machine vision system in these transplanters can classify seedlings based on factors such as color, volume, height, stem thickness, and root development. Currently, the FlexPlanter and MidiVision transplanters have been commercialized, becoming one of the more mature automated transplanting solutions on the market. The widespread application of these devices has not only improved agricultural production efficiency but also promoted the development of smart agriculture.

4.2.2 Research Status of Intelligent Identification of Plug Seedlings in China

In recent years, with the continuous advancement of agricultural automation technology, intelligent identification of plug seedlings has gradually become a research focus. In 2009, Yufeng Yu and his team at Jiangsu University designed a parallel transplanting robot based on a machine vision system. This system converted the pixel coordinates of each seedling's center into three-dimensional coordinates in the world coordinate system, enabling the actuating mechanism to precisely position the manipulator platform above the seedlings for grabbing.

This technology significantly improved the accuracy and efficiency of the transplanting process, marking an important application of machine vision technology in the agricultural sector. Subsequently, from 2013 to 2015, Jianmin Wu et al. at China Agricultural University further developed an automatic plug seedling identification and control device, which used photoelectric sensors for seedling recognition and could switch between manual and automatic control modes. Transplanting experiments demonstrated that the device effectively identified whether there were missing seedlings in the plug cells, significantly improving accuracy and efficiency.

At the same time, other studies have made progress. In 2013, Qingqing Feng et al. from Jiangsu University designed a plug seedling shape parameter measurement system based on structured light vision. This system captured image information of plug seedlings, achieving online measurements of leaf area and height. The experiments showed that the system's height measurement accuracy was 5 mm, and the leaf area measurement accuracy met screening requirements, enabling simple sorting of plug seedlings. In 2018, Yongwei Wang et al. from Zhejiang University developed an automatic seedling replenishment testbed for vegetable plug seedlings, and designed a corresponding machine vision system that accurately determined the boundaries of seedling holes and seedling information through color image processing. Subsequently, in 2019, Xin Zhou et al. from Jiangsu University developed an integrated device for sorting, transplanting, and replenishing plug seedlings, designing an intelligent detection scheme based on an RGB-D camera capable of processing 5,000 seedlings per hour.

To summarize, the current machine vision technologies in transplanters primarily employ two methods: one involves sensor-based detection, which can only detect missing seedlings and is significantly affected by the size of the plants, with relatively low intelligence levels. The other method is based on image processing, which extracts various features of plug seedlings such as leaf area and height through image segmentation algorithms. However, this method's effectiveness can be influenced by lighting conditions and camera angles, potentially leading to feature loss.

4.2.3 Research Status of Transplanting Technology Abroad

Jorg O. J. and his team designed two types of grippers, as shown in Fig. 4.1. One gripper functions by grasping the stem to pick the seedlings, while the other achieves this by inserting the gripper into the substrate to lift the seedlings with the soil. Transplanting experiments revealed that the average success rate for the two-finger gripper was 95%, while the needle-type gripper achieved a success rate of 81.75%.

Choi et al. designed a seedling picking mechanism using a five-bar linkage system, as illustrated in Fig. 4.2. This mechanism consists of a path generator, pick-up pins, and a pin driver to carry out seedling picking operations. Testing of seedling age, pin insertion depth, and gripping speed showed that insertion depth and gripping speed significantly affect seedling picking success rates. Using 23-day-old seedlings for transplanting tests, the mechanism picked 30 seedlings per minute with a success rate of 97%, meeting the design criteria for planting speed.

Rahul and his team designed another five-bar seedling picking mechanism, distinct in its possession of two degrees of freedom, as shown in Fig. 4.3. This mechanism is simple and controlled by a microcontroller. It offers good positional repeatability during the picking process without causing collisions or damaging the seedlings. However, due to its simplicity, it is not suitable for high-speed transplanting operations.

In summary, research on transplanting mechanisms abroad shows a diverse range of approaches. Different scholars have focused on improving the precision, automation, and adaptability of transplanting devices. These advancements are of great

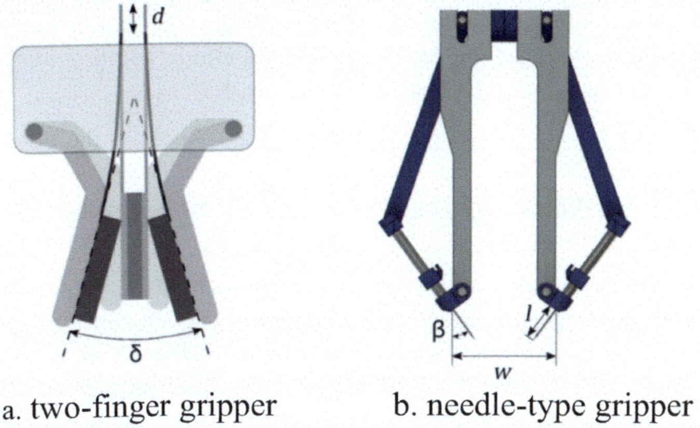

Fig. 4.1 Gripper types

Fig. 4.2 Five-bar mechanism for seedling picking

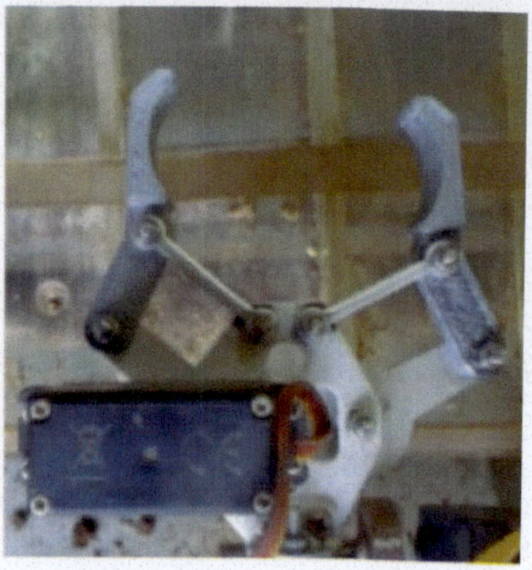

Fig. 4.3 Two-degree-of-freedom five-bar seedling picking mechanism

significance for agricultural production and horticulture, contributing to the modernization and sustainable development of global agriculture, and providing valuable references for our own research on seedling picking devices.

4.2.4 Research Status of Transplanting Technology in China

In China, transplanting machines are generally semi-automatic, requiring operator involvement during transplanting. With the continuous development of science and technology, and the push for agricultural modernization, there is an urgent need for a fully automatic transplanter in China to carry out transplanting operations. The seedling picking mechanism, as the key component of a fully automatic transplanter, has attracted significant research attention. Below are some studies conducted by Chinese scholars on seedling picking mechanisms.

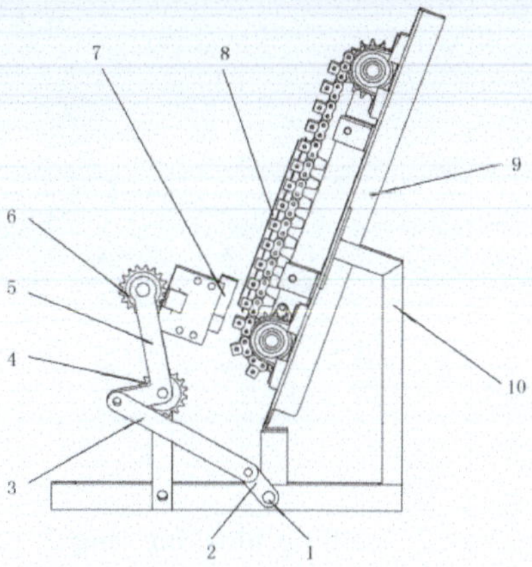

1. active shaft 2. crank 3. connecting rod 4. fixed sprocket
5. swing rod 6. moving sprocket 7. end effectort
8. seedling tray 9. seedling tray displacement mechanism 10. frame

Fig. 4.4 Schematic diagram of the swing rod seedling picking mechanism

Xianwei Hu and his team developed a seedling picking mechanism that operates on a swing rod principle, as shown in Fig. 4.4. After the gripper grasps the plug seedling, the drive shaft provides power to the device, and the motion of the crank-link mechanism drives the sprocket rotation, completing the seedling extraction and moving to the preset planting location. Test results showed that this picking mechanism achieved an average success rate of 97.1%.

Wei Cui and his team designed a second-order elliptical gear seedling picking mechanism, with the structure shown in Fig. 4.5. This mechanism uses a second-order elliptical gear transmission to achieve rotational motion. When the seedling claws reach the designated position, they extend forward and clamp the substrate. The claws then move the seedling to the planting position. When the picking cam reaches its lowest point, the claws retract to their initial position under spring tension. Finally, the seedling is released from the claws with the help of a seedling stopper ring, completing the transplanting process. Tests demonstrated that the success rate of this mechanism reached 91.3%.

Sibo Yao and his team designed a combined vibration-clamping seedling picking mechanism, shown in Fig. 4.6. This mechanism first loosens the plug seedling with vibration before using a clamping device to pick the seedling. Dynamics and vibration models revealed that the success rate of seedling picking depends on vibration frequency and substrate moisture content.

Jianping Hu and his team designed a top-clamp-pull seedling picking device, as shown in Fig. 4.7. The operation process is as follows: when the seedling claw moves above the plug tray, it grabs the plug seedling with the aid of a cylinder, while simultaneously pushing the substrate out using a top rod. The seedling is then

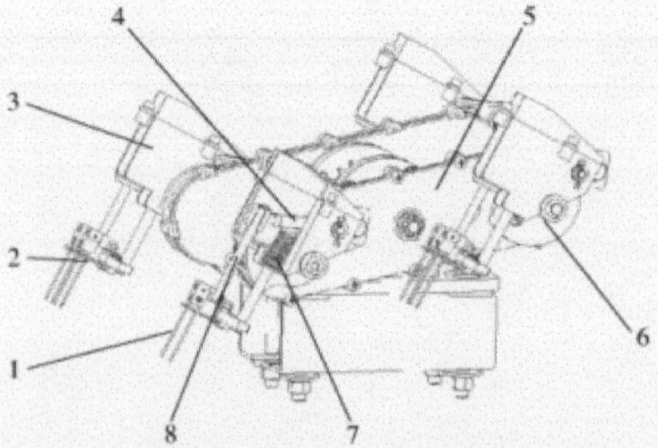

1. seedling claw 2. seedling blocking ring 3. seedling claw box
4. pressure rod 5. second order ellipticagearbox
6. seedling cam 7. spring 8. seedling claw push rod

Fig. 4.5 Structure of the second-order elliptical gear seedling picking mechanism

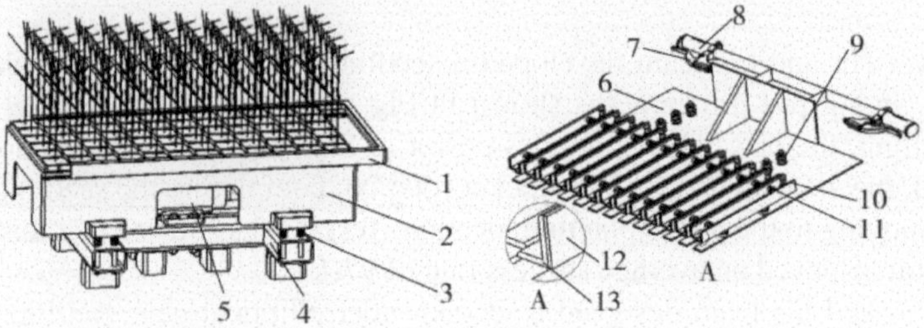

1. upper cover of vibration mechanism 2. vibration frame 3. vibration spring 4. frame 5. vibration motor 6. seedling picking mechanism frame 7. seedling picking control handle 8. anti slip sleeve 9. steel wire soft rope transmission roller 10. fine adjustment bolt 11. seedling picking spring 12. clamping protective sponge 13. clamping plate

Fig. 4.6 Combined vibration-clamping seedling picking mechanism

pulled out by the claw. The optimal operating conditions were found at a top rod displacement of 15 mm, an insertion depth of 35 mm, and a picking speed of 225 mm/s.

These seedling picking mechanisms all have their unique features and are capable of performing the seedling picking operation effectively. However, due to differences in agronomy, substrate types, and the uncertain growth status of crops in China's plug seedling cultivation, the studied mechanisms cannot yet be widely

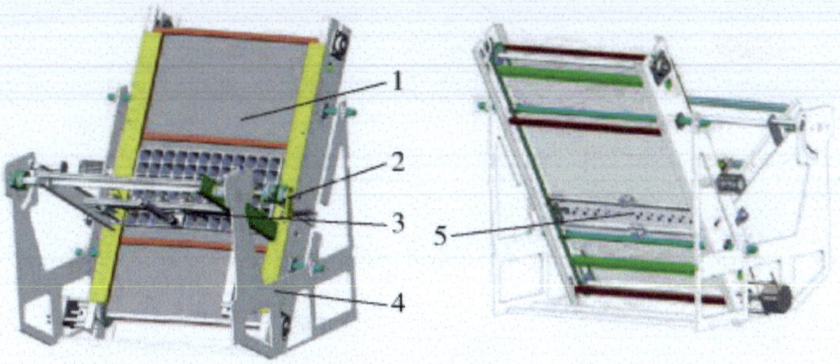

1. seedling tray conveying mechanism 2. seedling tray pressure rod mechanism 3. seedling picking mechanism 4. bracket 5. top seedling mechanism

Fig. 4.7 Top-clamp-pull seedling picking device

implemented on a large scale. Therefore, the development of seedling picking mechanisms must consider the local agricultural conditions and crop varieties to ensure their adaptability.

4.3 Research on Intelligent Identification Technology for Plug Seedlings

4.3.1 YOLOv5 Object Detection Network for Plug Seedlings

YOLOv5 is an object detection algorithm based on deep learning, with its name originating from "You Only Look Once (YOLO)." YOLOv5 builds upon previous versions such as YOLOv3 and YOLOv4, featuring an updated architecture. The YOLOv5 model comprises three parts: the backbone network, the head network, and the loss function. The network structure of YOLOv5 is shown in Fig. 4.8. Based on the number of layers and feature map channels, YOLOv5 can be divided into four models, from small to large: YOLOv5s, YOLOv5m, YOLOv5l, and YOLOv5x. The model used in this chapter is YOLOv5s.

The YOLOv5 network consists of three components: Backbone, Neck, and Head. Backbone: The task of the backbone is to extract features from the input image and gradually enhance the representation of the feature map. The backbone uses a CSPDarknet53 architecture, consisting of FOCUS modules, CBL modules, and SPP modules. The essence of the FOCUS module is a slicing operation, which is shown in Fig. 4.9. The CBL module is responsible for feature extraction and consists of three operations: CONV, BN, and LeakyReLU. This module helps improve the effectiveness of feature extraction. When feature maps of varying sizes are input, the SPP module is used to convert them into fixed-size features, as shown in Fig. 4.10.

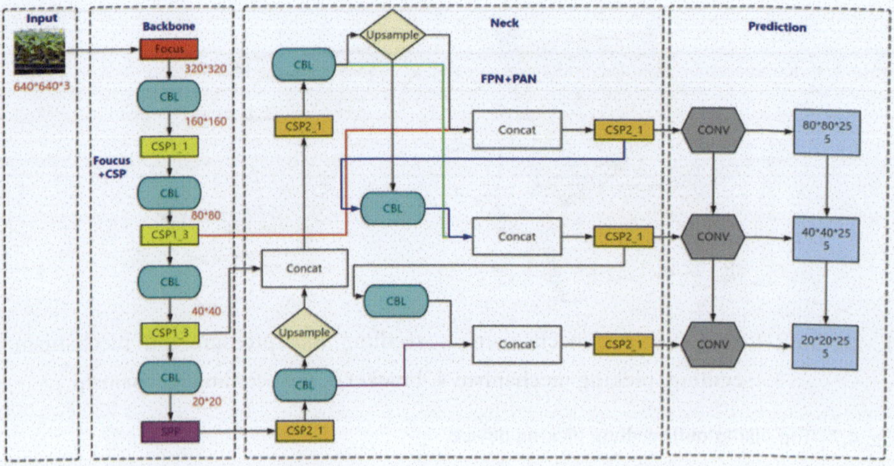

Fig. 4.8 Network structure of YOLOv5

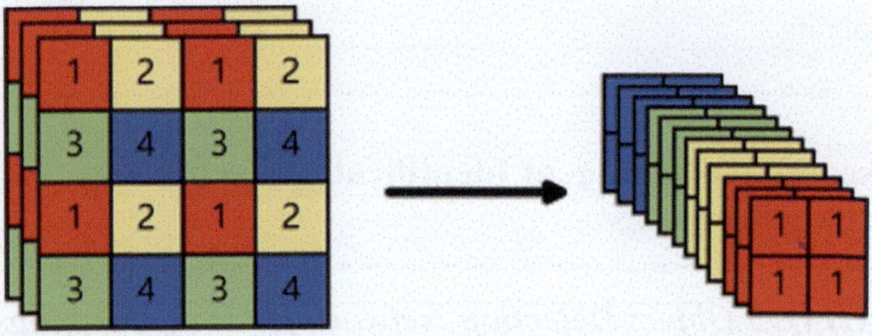

Fig. 4.9 Slicing operation

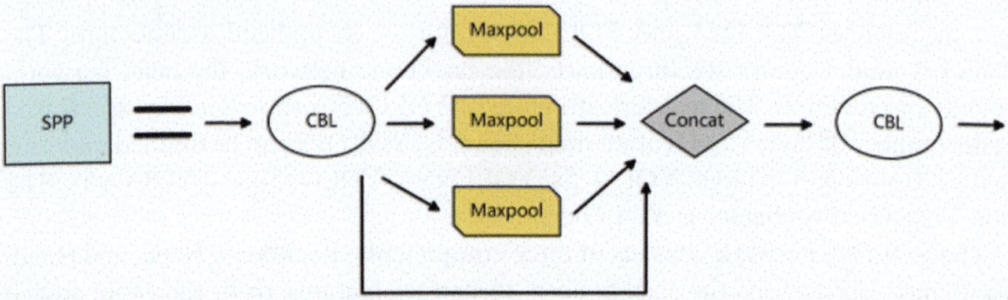

Fig. 4.10 Structure of the SPP module

Neck: Based on the backbone, the Neck adds a simple yet effective feature pyramid module. These modules focus on different areas of features, enhancing the quality and accuracy of feature representation. YOLOv5 adopts the FPN + PAN structure for the Neck.

Head: The Head is composed of three convolutional layers and is responsible for outputting the results of object detection, including object categories, bounding box positions, and confidence scores. The Head operates in three steps: feature extraction, object detection, and post-processing. YOLOv5 extracts features at different scales and generates a high-dimensional feature map to represent the input image. To adapt to objects of different scales, YOLOv5 uses feature pyramid modules with multiple branches, each with different receptive field sizes, to capture features of objects of varying sizes. Finally, the Head outputs the detection results, including the categories, confidence scores, and bounding box positions.

YOLOv5 uses post-processing techniques such as non-maximum suppression (NMS) and confidence thresholding to remove redundant bounding boxes from the detection results and output the final object detection outcome. Through this process, YOLOv5 can accurately recognize and locate objects in the input image.

4.3.2 Dataset Creation and Model Training

1. Dataset Creatio

Object detection algorithms generally require a large dataset to support them, and the detection performance depends on the quality of the dataset. Thus, building a dataset is a crucial step in object detection. To train and evaluate the performance of the YOLOv5 algorithm, a tomato seedling stem dataset was created. This section describes the collection of images of tomato plug seedlings, data augmentation processing, and data labeling.

Tomato plug seedling images were captured in a greenhouse seedling nursery. A total of 1,200 images of tomato plug seedlings were taken, and after organizing and selecting these images, low-quality ones were discarded. The remaining tomato plug seedling images are shown in Fig. 4.11.

Fig. 4.11 Tomato plug seedlings

To improve the accuracy and precision of tomato stem detection, data augmentation techniques such as Gaussian noise, salt-and-pepper noise, darkening, and brightness transformation were applied to the images. Data augmentation increases the diversity of the training data, allowing the model to better adapt to different scenarios and varying conditions in object recognition tasks. After processing, the images appear as shown in Fig. 4.12.

After augmentation, the total number of tomato plug seedling images reached 2,400. Of these, 800 images were selected as the validation set for the final evaluation of the model's performance, and the remaining 1,600 images were used as the training set. The purpose of data augmentation was to increase the number of images in the dataset. As the dataset size increased, the accuracy, robustness, and generalization capability of the YOLOv5 object detection algorithm improved, enabling the model to better learn object features in complex environments.

Before detection, the tomato plug seedling images were labeled. The labeling was performed using LabelImg to mark the tomato seedling stems in the images, and the labels were saved in a .txt format. Rectangular boxes were used for image labeling, ensuring that the seedling stems were centered within the box to reduce interference

c. Lighting and shade d. Brightness

Fig. 4.12 Images after data augmentation

Fig. 4.13 LabelImg labeling process

from extraneous background. The box positions were determined by the coordinates of two opposite corners. The labeling process using LabelImg is shown in Fig. 4.13.

2. Model Training

Model training was conducted in a lab environment. The experimental platform consisted of a Windows 11 operating system, a 12th Gen Intel(R) Core(TM) i9-12900KF processor, and the Pytorch deep learning framework. The development environment was PyCharm 2022, using Python 3.9.7 as the programming language, with Anaconda3 as the package manager.

The training settings for the model were as follows: input image size was set to 640 × 640, batch size was 16, initial learning rate was 0.01, and the number of training epochs was 200.

4.3.3 Improvements to the YOLOv5 Algorithm

To improve the accuracy of tomato seedling stem recognition and reduce the rate of missed detections, improvements were made to the original YOLOv5 algorithm. The enhancements include the addition of the lightweight upsampling operator CARAFE and the optimization of the loss function.

1. Addition of the lightweight upsampling operator CARAFE

In YOLOv5, upsampling refers to enlarging the size of the original image or feature map. Typically, YOLOv5 uses bilinear interpolation for upsampling, achieved with the F.interpolate function. Bilinear interpolation is a simple and effective method, but because it uses a small receptive field, it can limit performance. Therefore, to further enhance the performance of YOLOv5, the lightweight upsampling operator

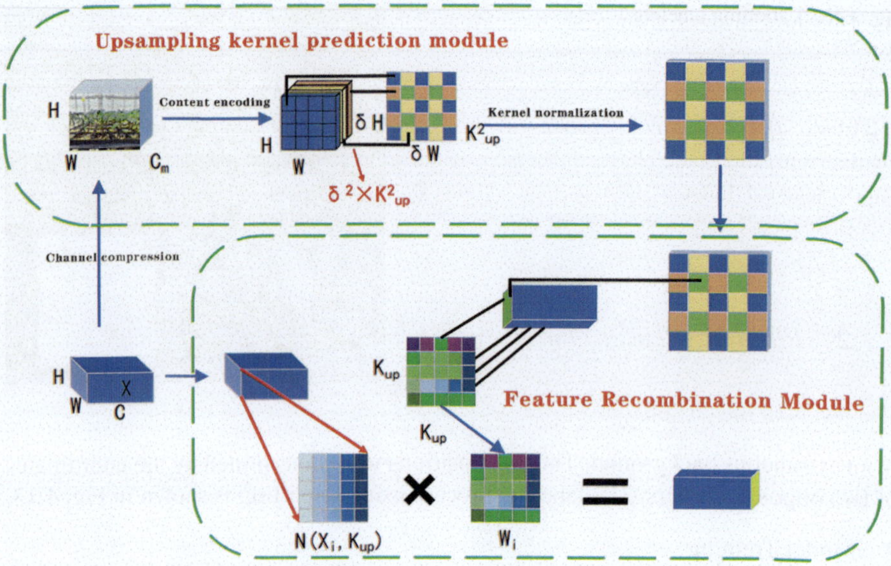

Fig. 4.14 CARAFE framework diagram

CARAFE was added to the YOLOv5s model. CARAFE can significantly increase the resolution and quality of the feature map, improving the model's performance while maintaining a lower computational load compared to traditional upsampling operators. It is composed of two main modules: the kernel prediction module and the feature reassembly module, as shown in Fig. 4.14.

The kernel prediction module predicts a reassembly kernel for each target location based on the content at that location, enabling adaptive convolution operations. The process involves compressing the channel dimension of the input feature map (i.e., performing slicing), predicting a reassembly kernel for each target location using a local attention mechanism, and performing adaptive convolution operations. The feature reassembly module uses the predicted kernels to reassemble the feature maps, generating more distinguishable features. The process involves: (1) Using the predicted kernels to perform adaptive convolution on the input feature map, generating a feature representation for each target location. (2) Merging the generated feature representations with the original features. (3) Applying a convolution layer to the merged features, followed by the output of the processed features.

Compared to bilinear interpolation, CARAFE has a larger receptive field and can aggregate contextual information over a wider area. Additionally, CARAFE supports instance-specific content-aware processing, dynamically generating adaptive kernels. When integrated into the Feature Pyramid Network (FPN) structure of YOLOv5, CARAFE allows for better upsampling and significantly outperforms the original operator. By incorporating CARAFE into the network, the model's performance is effectively improved without significantly increasing its computational complexity.

2. Improvement to the loss function

The original YOLOv5s model used CIoU (Complete Intersection over Union) as its loss function. The improvement made here involves replacing CIoU with EIoU (Extended Intersection over Union).

CIoU is an advanced metric for measuring the distance between two bounding boxes, improving upon traditional IoU by considering additional factors such as aspect ratio. The CIoU formula is as follows:

$$L_{CIoU} = 1 - IoU + \frac{\rho^2(b, b^{gt})}{c^2} + \alpha V \tag{4.1}$$

$$V = \frac{4}{\pi^2}\left(arctan\frac{w^2}{h^{gt}} - arctan\frac{w}{h}\right)^2 \tag{4.2}$$

$$\alpha = \begin{cases} 0, & \text{if } IoU < 0.5 \\ \frac{V}{(1-IoU)+V}, & \text{if } IoU \geq 0.5 \end{cases} \tag{4.3}$$

where V represents the normalized aspect ratio difference between the predicted and ground truth boxes, b is the parameter for predicting the center coordinates, b^{gt} is the parameter for the center of the real target bounding box, c is the diagonal length of the smallest enclosing box, α is a balancing factor for the aspect ratio loss and IoU loss, ρ denotes the Euclidean distance between the centers of the predicted and ground truth boxes, w, h and w^{gt}, h^{gt} are the height and width of the predicted box and the height and width of the real box, respectively.

While CIoU improves upon DIoU (Distance IoU) by adding a scale-based loss term, it still has limitations. Specifically, it does not accurately reflect the relationship between the width and height of the bounding box and the confidence score. As a result, the CIoU loss function can hinder effective optimization in certain situations.

To address this issue, EIoU was proposed in 2021. EIoU further decomposes and optimizes the aspect ratio, while incorporating Focal Loss techniques to enhance the quality of anchor boxes. The formula for EIoU is as follows:

$$\begin{aligned} L_{EIoU} &= L_{IoU} + L_{dis} + L_{asp} \\ &= 1 - IoU + \frac{\rho^2(b, b^{gt})}{c^2} + \frac{\rho^2(w, w^{gt})}{c_h^2} + \frac{\rho^2(h, h^{gt})}{c_h^2} \end{aligned} \tag{4.4}$$

where C_w and C_h are the width and height of the smallest enclosing box. EIoU consists of three components: overlap loss, center distance loss, and width-height loss. It retains the overlap area and center distance from CIoU but introduces optimization to the width and height losses, accelerating convergence. Additionally, the integration of Focal Loss addresses the imbalance of training samples, further improving accuracy. EIoU not only speeds up convergence but also improves the accuracy of bounding box regression. Moreover, EIoU addresses the limitation of CIoU, where width and height cannot be adjusted simultaneously, further enhancing the performance of object detection models.

4.3.4 Comparative Analysis Before and After Algorithm Improvements

1. Model evaluation metrics

The detection results of YOLOv5 can be represented by several metrics. Precision and Recall are two important indicators for evaluating the performance of YOLOv5's object detection system. Precision refers to the proportion of correctly detected objects among all detected objects, and recall represents the probability that real objects were correctly detected. Higher precision and recall values indicate better system performance. The metrics used to evaluate the performance of the model include mean Average Precision (mAP), Precision (P), Recall (R), and Accuracy (A). The formulas are as follows:

① Precision (P)

$$P = \frac{TP}{TP + FP} \tag{4.5}$$

where TP represents true positives (correctly classified positive instances), and FP represents false positives (incorrectly classified negative instances).

② Recall (R)

$$R = \frac{TP}{TP + FN} \tag{4.6}$$

where FN represents false negatives (incorrectly classified positive instances).

③ Accuracy (A)

$$A = \frac{TP + TN}{TP + EP + FN + TN} \tag{4.7}$$

where TN represents true negatives (correctly classified negative instances).

④ Mean Average Precision (mAP).

mAP is a comprehensive performance evaluation metric that represents the average precision across all categories within an image. The higher the mAP, the higher the prediction accuracy of the model across different categories.

2. Ablation experiment

In the tomato seedling stem recognition experiment, two modules of the YOLOv5s algorithm were improved. To test the optimization effects of each improvement, ablation experiments were designed based on these two modules, with YOLOv5s serving as the baseline. The results are shown in Table 4.1, where "√" indicates that the module was improved, and "×" indicates no improvement.

The table shows that the improvements made to the original algorithm led to performance gains in each case. Adding the lightweight upsampling operator CARAFE resulted in more effective feature extraction and a reduction in model size and detection time. The loss function modification with EIoU also improved performance, although to a lesser degree than CARAFE. The combined improvements of CARAFE and EIoU provided the best results, achieving a 95.4% mAP, an 11 ms reduction in average detection time, and a 2.09 MB decrease in model size compared to the original YOLOv5s.

3. Performance comparison before and after the improvements

The improvements to YOLOv5s through the addition of CARAFE and EIoU were evaluated by running detection tasks on the tomato seedling stem dataset, which comprised 1,800 images. Key performance metrics, including recall, precision, and mAP, were compared between the improved and original YOLOv5s models. The results are summarized in Table 4.2.

From the table, it is clear that the improvements made to the YOLOv5s algorithm resulted in significant enhancements in performance. After the improvements, the recall increased to 94.1%, the precision reached 92.6%, and the mAP at 0.5 increased to 95.4%. These represent improvements of 3.3%, 3.3%, and 3.2%, respectively, compared to the original model. To provide a clearer illustration of these improvements, graphs comparing recall, precision, and mAP at 0.5 across the same number of iterations are shown in Fig. 4.15.

Table 4.1 Ablation experiment results

Method	CARAFE	EIOU	mAP_0.5 (%)	Average detection time (ms)	Model size (MB)
YOLOv5s	×	×	92.2	47	6.35
YOLOv5s + EIOU	×	√	93.2	43	4.93
YOLOv5s + CARAFE	√	×	93.6	38	4.21
Improved algorithm	√	√	95.4	36	4.14

Table 4.2 Performance metrics

Algorithm	Recall (%)	Precision (%)	mAP_0.5 (%)
YOLOv5	90.8	89.3	92.2
Improved YOLOv5s	94.1	92.6	95.4

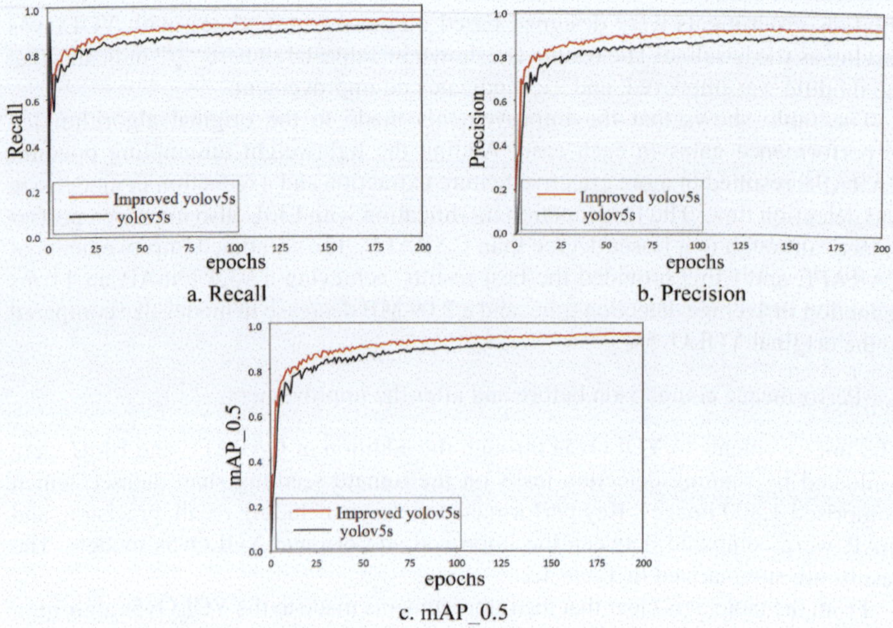

Fig. 4.15 Performance comparison

To evaluate the performance of the improved YOLOv5s model on new, unseen images, two images that were not used in the network training process were tested. The results are shown in Fig. 4.16.

From Fig. 4.17, it is evident that the improved model achieved better detection results. Compared to the original algorithm, the confidence level of the detection results after improvement was higher, indicating more accurate identification of tomato seedling stems. This demonstrates that the improved algorithm effectively

a. Test result 1 b. Test result 2

Fig. 4.16 Detection results of the improved YOLOv5 model

c. Original results d. Improved results

Fig. 4.17 Test results before and after the algorithm improvements

resolved the issue of low detection precision for tomato seedling stems, resulting in better overall detection performance.

4.3.5 Analysis of Tomato Seedling Stem Detection

During the transplanting process of tomato plug seedlings, variations in the growth conditions of the seedlings lead to differences in stem appearance. Seedlings in good growing conditions typically have straight or slightly curved stems. However, during the seedling growth process, some poorly growing seedlings inevitably emerge, which, if transplanted, can affect the survival rate of the tomato plants. Therefore, poorly growing seedlings should be excluded during the transplanting process. The appearance characteristics of poorly growing seedlings typically involve severely curved stems, as shown in Fig. 4.18.

Fig. 4.18 Severely bent stems

Fig. 4.19 Stem straightness range

The degree of bending of the tomato seedling stem is referred to as stem straightness, defined as the angle β between the main stem of the tomato seedling and the horizontal direction of the plug tray. The evaluation criteria for seedling straightness are $45° < \beta < 135°$. When β falls within this range, the tomato seedling is suitable for transplanting. The range of acceptable stem straightness is illustrated in Fig. 4.19.

Seedlings with good growth conditions were labeled, and the YOLOv5 algorithm was trained on these samples. After training, the model was tested and validated. When there were poorly growing tomato seedlings present in the plug tray, the detection results were as shown in Fig. 4.20. The results indicate that the algorithm successfully identified the healthy tomato seedling stems, while the stems of seedlings that did not meet the straightness criteria for transplanting were not marked, thereby achieving the goal of excluding substandard seedlings from transplanting.

a. Test result 1 b. Test result 2

Fig. 4.20 Test results

4.4 Research on Plug Seedling Positioning Technology

Positioning the tomato seedling stem is achieved through the use of a depth camera. Seedling positioning is a complex process that requires several steps. First, the camera model must be established, the imaging principles of the depth camera must be clarified, and the relationships between various coordinate systems must be converted. Then, the camera must be calibrated to obtain the necessary parameters, and finally, hand-eye calibration is conducted.

4.4.1 Camera Imaging Model

In the process of recognizing and grabbing objects, the depth camera is a key component, providing scene information to the system. The camera imaging model can be understood as a simplified optical model that describes the transformation from the world coordinate system to the pixel coordinate system.

(1) **World Coordinate System**: Describes the camera's position, denoted as (X_w, Y_w, Z_w);
(2) **Camera Coordinate System**: Describes the position of objects relative to the camera;
(3) **Image Physical Coordinate System**: Has its origin at the center of the CCD image plane, with x and y axes parallel to the pixel coordinate system, represented by (x, y);
(4) **Image Pixel Coordinate System**: Represents the projection of objects onto the image plane.

The transformation from world coordinates to pixel coordinates involves translation, rotation, scaling, and the application of the principle of similar triangles. The relationships between the four coordinate systems are shown in Fig. 4.21.

A camera serves as the medium linking the three-dimensional real world with the two-dimensional image plane. The most widely used camera imaging model is the pinhole imaging model, whose principle is illustrated in Fig. 4.22. This study utilizes an Intel RealSense D435i depth camera, which uses an RGB sensor on its far-right side. Therefore, the RGB camera operates based on the pinhole imaging model.

Let the coordinates of point p be (x, y, z), and the coordinates of its image p′ be (x', y', z'). Based on the principle of similar triangles, the following relationship is obtained:

$$\frac{d}{f} = -\frac{x}{x'} = -\frac{y}{y'} \tag{4.8}$$

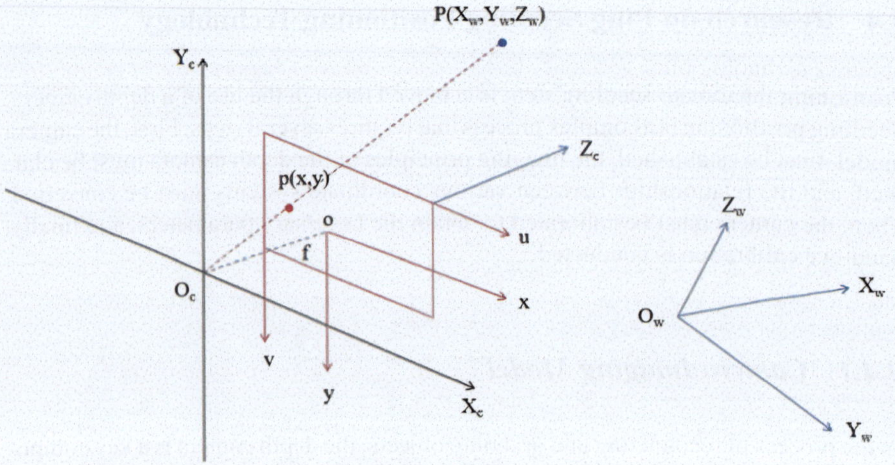

Fig. 4.21 Transformation between coordinate systems

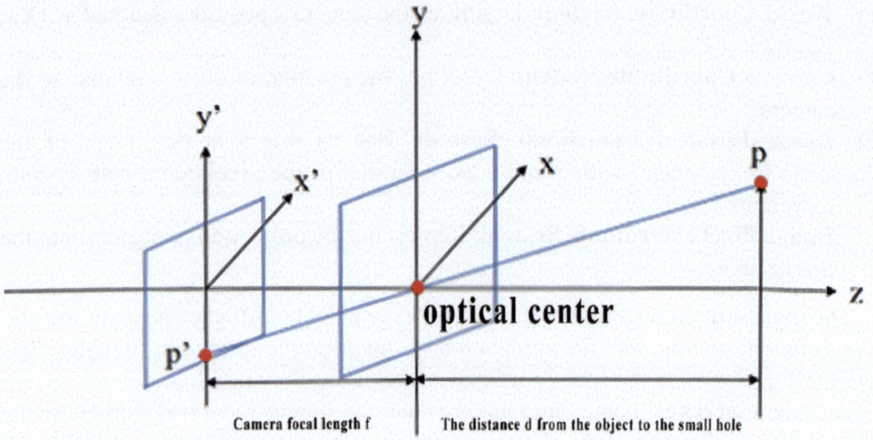

Fig. 4.22 Pinhole imaging principle

Since the image formed by the pinhole model is inverted, a negative sign is used. To simplify the equation, we can symmetrically flip the image plane and obtain the following:

$$\frac{d}{f} = \frac{x}{x'} = \frac{y}{y'} \tag{4.9}$$

Rearranging the equation yields:

$$x' = f\frac{x}{d} \tag{4.10}$$

$$y' = f\frac{y}{d} \tag{4.11}$$

At this point, (x', y') represents the coordinates in the image physical coordinate system. By scaling the coordinates along the u-axis by a factor of α and along the v-axis by a factor of β, and translating the origin by (Cx, Cy), the relationship between the coordinates of point p' and the pixel coordinates is as follows:

$$\begin{cases} u = \alpha x' + C_x \\ v = \beta y' + C_y \end{cases} \tag{4.12}$$

Rearranging the equations yields:

$$\begin{cases} u = f_x \frac{x}{d} + C_x \\ \\ v = f_y \frac{y}{d} + C_y \end{cases} \tag{4.13}$$

In matrix form:

$$d\begin{pmatrix} u \\ v \\ 1 \end{pmatrix} = \begin{pmatrix} f_x & 0 & C_x \\ 0 & f_y & C_y \\ 0 & 0 & 1 \end{pmatrix}\begin{pmatrix} x \\ y \\ z \end{pmatrix} \triangleq Kp \tag{4.14}$$

where K represents the camera's intrinsic matrix. This process describes pinhole imaging, which transforms world coordinates into pixel coordinates.

4.4.2 *Camera Calibration*

Camera calibration can be categorized into three types based on different principles:

(1) **Traditional Camera Calibration**: This method is applicable to any camera model, providing high accuracy. However, it always requires a calibration object and involves complex algorithms. Common methods include Tsai's two-step method and Zhang's calibration method.
(2) **Active Vision Camera Calibration**: Compared to traditional calibration methods, this method does not require a calibration object. It is simpler, more robust, and suitable for certain situations.
(3) **Self-Calibration**: This method can be used for online calibration, offering flexibility, but it has lower precision and robustness. Common self-calibration techniques include hierarchical step-by-step calibration and Kruppa equation-based methods.

Fig. 4.23 Calibration checkerboard

In this research, the Zhang Zhengyou calibration method was used to obtain camera parameters. This method involves using a checkerboard calibration board. The process consists of capturing multiple images of the calibration board from different angles, from which the image coordinates of the checkerboard's corner points are identified. By combining this information with the known physical coordinates of the board, the camera's intrinsic and extrinsic parameters, as well as lens distortion parameters, can be calculated. The checkerboard used in the experiment is shown in Fig. 4.23. This method involves using a checkerboard calibration board. The process consists of capturing multiple images of the calibration board from different angles, from which the image coordinates of the checkerboard's corner points are identified. By combining this information with the known physical coordinates of the board, the camera's intrinsic and extrinsic parameters, as well as lens distortion parameters, can be calculated. The checkerboard used in the experiment is shown in Fig. 4.23.

Radial Distortion: This type of distortion is common in camera lenses, with the distortion becoming more noticeable as points move further from the optical center. Near the optical center, radial distortion can be neglected, but towards the edges, it becomes more prominent. Radial distortion can be corrected using a Taylor series expansion, as shown below:

$$X_{corrected} = X(1 + K_1r^2 + K_2r^4 + K_3r^6) \tag{4.15}$$

$$Y_{corrected} = Y(1 + K_1r^2 + K_2r^4 + K_3r^6) \tag{4.16}$$

where X and Y represent the distorted pixel coordinates, while the corrected coordinates are obtained by compensating for radial distortion. The value r denotes the distance of a point from the image center.

4.4.3 *Calibration Experiment*

Camera calibration technology has matured over time, with popular methods including MATLAB and OpenCV calibration. In this experiment, MATLAB's calibration toolbox was used for camera calibration. The calibration was conducted using a 7 × 9 black-and-white checkerboard pattern. The checkerboard was printed and images were captured using a depth camera. A total of 12 images of the calibration board were captured from various angles. MATLAB's Camera Calibration toolbox was then used to process these images, yielding the camera's intrinsic parameters and distortion coefficients. The results of the calibration experiment are shown in Fig. 4.24, with the intrinsic matrix parameters listed in Table 4.3.

4.5 Design of the Seedling Picking Manipulator

As a crucial tool in modern agricultural automation, the design and application of seedling picking manipulators cover several key areas. The primary goal of these manipulators is to replace manual labor by automatically grabbing, transplanting,

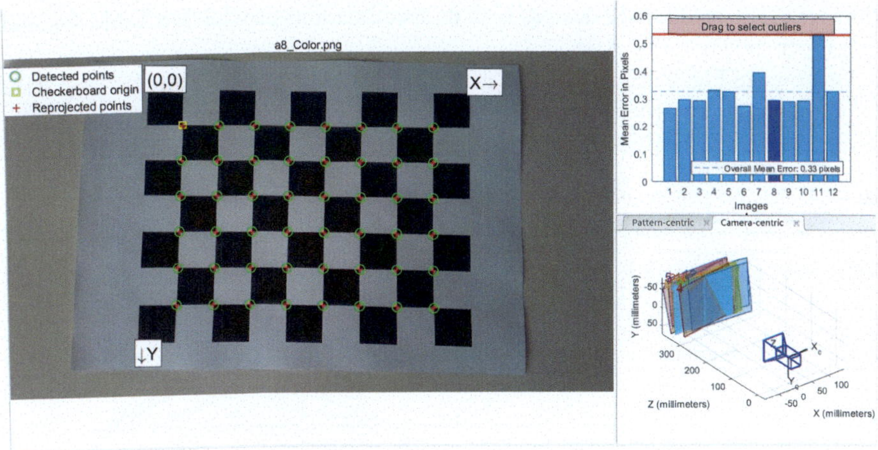

Fig. 4.24 Calibration experiment results

Table 4.3 Intrinsic matrix parameters

Parameter	Value
f_x	982
f_y	979
C_x	649
C_y	373

and placing seedlings. The manipulator typically consists of multiple joints, allowing it to simulate a variety of human hand movements, such as gripping, rotating, and moving, granting it a high degree of flexibility and precision. Among these, the design of the end-effector is especially important, as it directly contacts the seedlings during operation. Since seedlings are fragile, the end-effector must exert enough force to hold the seedling without causing damage during handling.

To meet this requirement, researchers have developed various types of end-effectors, including vacuum suction cups, mechanical grippers, and flexible gripping devices. Each type has specific advantages and is suited to different scenarios. Through these end-effectors, manipulators can perform highly accurate and efficient seedling picking tasks, greatly improving the level of agricultural automation.

In the case of seedling picking, the selection and design of the mechanical gripper are crucial, as the performance of the gripper directly impacts the efficiency and accuracy of transplanting operations. Several factors must be considered, including the gripping range, gripping method, material, and durability. Based on these factors, and in combination with the design requirements, a two-finger gripper was selected. To minimize damage to the plug seedlings during handling, the gripper's fingertips were made from flexible materials.

The seedling picking process requires precise control and high stability, with the gripper needing to withstand prolonged use. Given these demands, a servo motor-driven flexible gripper was chosen. In particular, the WHEELTEC two-finger flexible gripper was selected for the design, with its specifications provided in Table 4.4, and the gripper shown in Fig. 4.25.

As shown in Fig. 4.25, the flexible gripper operates by gripping the seedling stem at the fingertips. To avoid interference with neighboring seedlings in the plug tray,

Table 4.4 Flexible gripper specifications

Driving method	Gripping weight (kg)	Voltage (V)	Gripping frequency (times/min)
Servomotor	0.5	5	<60

Fig. 4.25 Flexible gripper

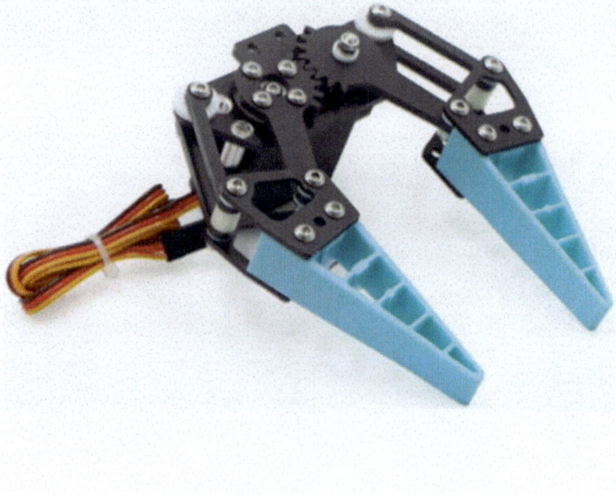

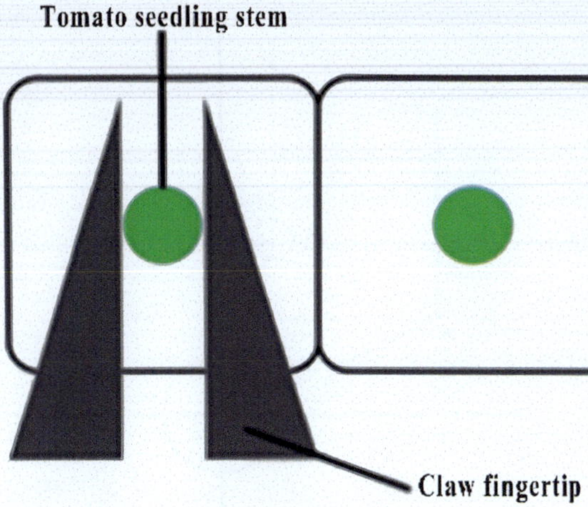

Fig. 4.26 Gripping process of the gripper

the gripper's fingertip length must be smaller than the diameter of an individual plug cell. The plug tray used in this research was a 6 × 12 tray, with a top opening size of 38 × 38 mm for each cell. Therefore, the fingertip length of the gripper should not exceed 38 mm (Fig. 4.26).

Additionally, the gripper operates through a parallel opening and closing mechanism. As a result, the opening distance of the gripper must also be considered to avoid exceeding the dimensions of the plug cell. The gripping diagram is shown in Fig. 4.27.

The gripper opening distance is represented as 2ΔH, where ΔH is half of the gripper's opening distance in millimeters:

$$\Delta H = H_2 - H_1 \tag{4.17}$$

where H_1 is the gripper's closed position distance, mm; H_2 is the gripper's open position distance, mm.

$$H_1 = R \times \sin\alpha \tag{4.18}$$

$$H_2 = R \times \sin\beta \tag{4.19}$$

where R is the distance from point O to A (or A'), mm; α is the gripper's closed position angle, °; β is the gripper's open position angle, °.

Given the values of R, α and ΔH, the corresponding β for suitable seedling picking was determined to be 22°.

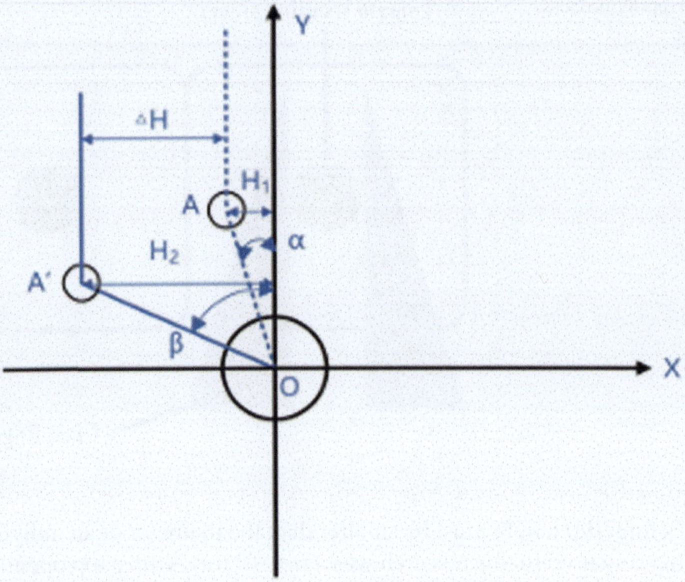

Fig. 4.27 Gripper gripping diagram

4.6 Conclusion of Research on Intelligent Identification and Transplanting Technology for Plug Seedlings in Desert and Gobi Regions

This study addresses the challenges of harsh environmental conditions in desert and Gobi regions by developing and validating an intelligent identification and transplanting technology for plug seedlings. Using the YOLOv5 object detection network, the research team successfully achieved accurate identification of tomato seedling stems. The creation of a dataset and model training, along with key improvements to the YOLOv5 algorithm, such as adding the lightweight upsampling operator CARAFE and optimizing the loss function from CIoU to EIoU, resulted in significant improvements in recognition accuracy. Experimental results demonstrated that the improved algorithm outperformed the original in all metrics, particularly in terms of tomato seedling stem recognition accuracy, achieving a 3.2% improvement. The recall and precision rates were also increased by 3.3% each. This high-accuracy recognition technology lays a strong foundation for subsequent automation in transplanting operations, ensuring that poorly growing seedlings are effectively filtered out, ultimately improving the overall transplant success rate.

In terms of transplanting technology, several seedling picking mechanisms were designed, including flexible grippers, vibration-assisted picking devices, and top-clamp-pull mechanisms. Detailed performance tests and comparative analyses were conducted. These innovative designs effectively address the challenges faced by traditional transplanters operating in desert and Gobi areas, such as high seedling damage rates and low picking success rates. In particular, the design of the flexible gripper took into account the growth characteristics of plug seedlings. By using a servo motor to control the gripper's opening and closing, the gripper was able to securely grasp the stem without damaging the seedling. Experimental data showed that the flexible gripper achieved a 97.1% success rate in picking seedlings, improving both the efficiency and accuracy of automated transplanting. The integrated application of this series of technologies has successfully achieved the organic combination of intelligent recognition and precise transplantation, bringing new hope to agricultural production in desert areas. It not only reduces labor input, but also improves the overall efficiency and sustainability of agricultural production. In the future, with further optimization and promotion of technology, this research achievement is expected to play an important role in a wider range of agricultural fields, providing strong support for global agricultural modernization and intelligence.

Chapter 5
Design and Experiment of Intelligent Transplantation Device for Desert Gobi Cave Seedlings

Can Hu, Long Wang, Zhicheng Liu, Boxuan Gu, and Weihua Cao

5.1 Introduction

Southern Xinjiang is an important agricultural production area in China. However, due to the unique terrain and dry climate in southern Xinjiang, especially in winter, the precipitation is scarce and the temperature is low, which has brought great trouble to agricultural production. The traditional agricultural planting mode is difficult to achieve high yield and high efficiency under such climatic conditions, and there are also many problems in the utilization of land and water resources. Therefore, in order to solve these problems and improve the yield and quality of agricultural products, the southern Xinjiang region began to actively promote the development of facility agriculture. Facility agriculture is a way to use facilities (such as greenhouses, greenhouses, etc.) to create a suitable growth environment for planting crops. Compared with traditional open field planting, facility agriculture has many advantages, such

C. Hu · L. Wang (✉) · Z. Liu · B. Gu
College of Mechanical and Electrical Engineering, Tarim University, Alar 843300, China
e-mail: 120140002@taru.edu

Modern Agricultural Engineering Key Laboratory at Universities of Education Department of Xinjiang Uygur Autonomous Region, Alar 843300, China

Xinjiang Production and Construction Corps (XPCC) Key Laboratory of Utilization and Equipment of Special Agricultural and Forestry Products in Southern Xinjiang, Alar 843300, China

Z. Liu
e-mail: 10757222241@stumail.taru.edu.cn

W. Cao
Key Laboratory of Smart Agriculture System Integration, Ministry of Education, Beijing 100083, China

Key Laboratory of Agricultural Information Acquisition Technology, Ministry of Agriculture and Rural Affairs, Beijing 100083, China

C. Hu et al. (eds.), *New Technologies Applied in Gobi Desert Facility Agriculture*, Smart Agriculture 11, https://doi.org/10.1007/978-981-96-1031-0_5

as providing a stable temperature and humidity environment, ensuring the normal growth and development of vegetables, preventing wind and sand, and reducing water evaporation. Facility agriculture marks the mechanization and intellectualization of agricultural production. With the rapid development of the national economy and the improvement of science and technology, facility agriculture will be gradually popularized and promoted. The development of facility agriculture is inseparable from the support of related technical equipment and machinery. At present, the demand for facility machinery in southern Xinjiang is increasing year by year.

As of 2022, China's facility agriculture encompasses an area of 2.85 million hectares, and China stands as the country boasting the largest area of facility agriculture. Over the past five years, the planting area of facility agriculture in China is depicted in Fig. 5.1. It can be distinctly observed that the area of facility agriculture in China has been on the rise over the past three years. Among greenhouse vegetables, the cultivation of greenhouse tomatoes is relatively straightforward and features high yield and income, thus it gradually assumes an important position in the production of greenhouse vegetables in southern Xinjiang. In recent years, the area of facility tomato planting in southern Xinjiang has been expanding. At present, it mainly relies on manual operation and lacks the application of facility mechanization, which limits the improvement of agricultural production efficiency and quality, and cannot meet the urgent needs of the market for automatic planting of greenhouse vegetables. Therefore, the application of facility machinery to facility agriculture is of great significance for promoting the development of facility agriculture.

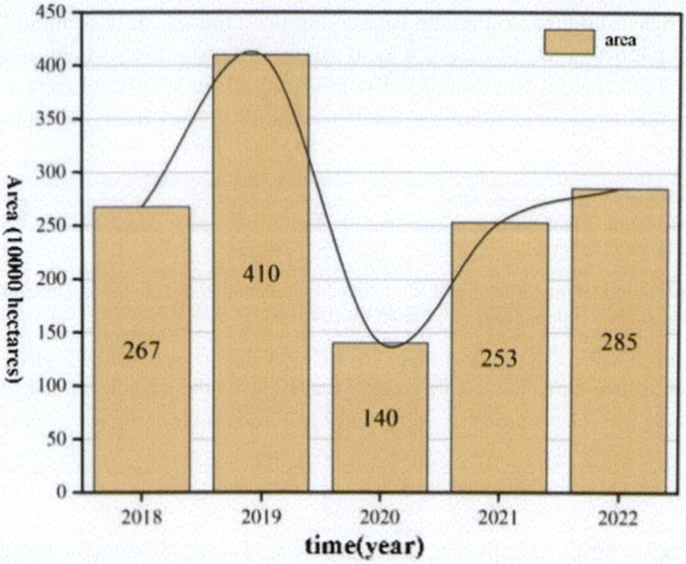

Fig. 5.1 Planting area of facility agriculture in China

Seedling transplanting is an indispensable part of tomato production. Seedling transplanting is not only a complex work, but also a high-intensity labor demand. The swift advancement of the vegetable sector is inherently intertwined with manual labor. However, as labor costs escalate, the adoption of mechanized production processes has emerged as a pivotal catalyst for the industry's rapid growth. Despite this, the mechanization of China's vegetable industry has been a relatively recent endeavor, with the machinery's efficiency often hindered by the diversity of crop varieties and the intricacies of planting patterns. Additionally, the technology for mechanized transplanting remains underdeveloped, and there is a notable deficiency in both the research and the widespread implementation of these technologies. When considering the overall mechanization landscape, it is evident that China's vegetable industry lags significantly behind the agricultural mechanization benchmarks set by developed nations. Currently, the cultivation and transplanting of vegetables in protected facilities in southern Xinjiang predominantly rely on manual labor, which is both labor-intensive and costly. This is particularly true for large-scale operations, where the workload is significantly higher. Moreover, manual transplanting demands meticulous management and operation, placing a premium on the technical proficiency of the farmers. Vegetable planting often emphasizes the importance of planting density and uniformity to optimize the use of limited space and resources. However, manual transplanting can lead to inconsistencies in planting spacing and an uneven distribution of plants. The facility vegetable transplanter can accurately control the planting spacing and depth to ensure the uniform arrangement and growth of the plants. In recent years, the rural population has gradually migrated to the city, and there are few people engaged in agricultural production, which leads to higher and higher labor costs, and artificial transplanting is increasingly unable to keep up with the pace of modern facility agriculture. Therefore, it is of great significance to study a facility plug seedling transplanter suitable for facility agriculture in southern Xinjiang to improve operation efficiency, reduce operation cost and realize facility automation.

Therefore, according to the characteristics of facility tomatoes in southern Xinjiang and the transplanting requirements of facility tomatoes, a seedling picking mechanism that can realize automatic seedling picking of facility transplanting machine is designed. The appropriate control mode and mechanical claw are selected to improve the operation accuracy and grasping efficiency during seedling picking and reduce the damage to tomato seedlings during seedling picking. It provides some reference for the design optimization of facility plug seedling transplanting machine.

5.2 Development Status of Facility Transplanting Machine

5.2.1 Current Situation of Overseas Research

Compared with the development of domestic facility transplanting machines, foreign research has been carried out earlier. The early transplanting devices are mainly industrial robots, supplemented by mechanical claws. In 1987, KUTZ designed a transplanting robot for seedbed plants. The transplanting machine uses a Puma 560 robot with 6 degrees of freedom as the main body. Through the research, it is found that the working condition of the transplanting robot is affected by the position of the hole tray. Finally, through experimental improvement, the transplanting robot transplanted 96% of the seedlings without causing damage to the seedlings. The average cycle time of transplanting seedlings of a 36-hole tray was 3.3 min.

In 1992, YANG, TING et al. studied a transplanting robot with SCARA industrial robot as the main body, as shown in Fig. 5.2. The end effector of the transplanter is a sliding needle holder with a sensor, through which the sensor can judge whether the pot seedling is caught and avoid the phenomenon of seedling damage when the holder is caught. At the same time, the factors affecting the operation of the gripper were studied, and the gripper was improved. Finally, the test was carried out to pick up the pot plate seedlings with different growth conditions. The test results showed that the transplanting success rate of the transplanter was over 50%.

Visser HortiSystems is a renowned company with an extensive history in the research and design of horticultural machinery. Their innovative pic-o-matvision transplanter is equipped with advanced capabilities, such as the ability to identify and eliminate holes and substandard seedlings. This ensures that every seedling transplanted by the machine meets the highest quality standards, with a remarkable 100% success rate. Moreover, the transplanting efficiency is impressive, with the machine capable of handling up to 2000 plants per hour. The transplanting machine is depicted in Fig. 5.3.

A company in the Netherlands developed the PC-21 automatic transplanter. The transplanter adopts advanced mechanical and automation technology, which can quickly and accurately identify and grasp the hole disc seedlings, and accurately locate and implant them into the preset culture medium, as shown in Fig. 5.4. The efficiency of this transplanting machine is remarkable, achieving a rate of 15,000 plants per hour, which is 15–20 times that of manual transplanting.

At present, foreign greenhouse transplanter technology is more mature, not only high efficiency but also can be combined with the vision system to eliminate holes, greatly saving labor. However, due to the large volume, high price and complex structure of foreign greenhouse transplanter, the foreign greenhouse transplanter is more suitable for factory seedling cultivation and is not suitable for facility agriculture in China.

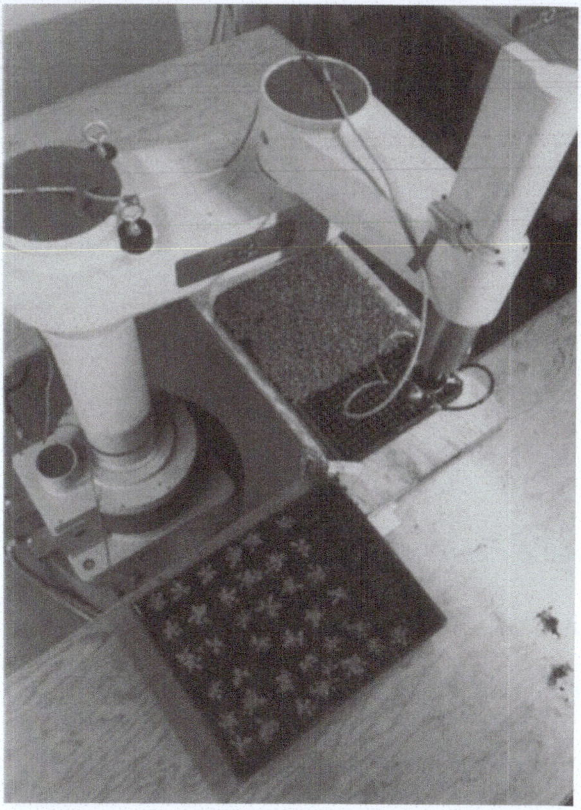

Fig. 5.2 SCARA transplanter

Fig. 5.3 pic-o-matvision transplanter

Fig. 5.4 PC-21 greenhouse pot plate seedling automatic transplanting machine

5.2.2 *Current Situation of Domestic Research*

In China, greenhouse planting is one of the main off-season vegetable production methods. Through greenhouse cultivation, farmers can grow and harvest vegetables during the unconventional season, thus meeting the market demand for fresh vegetables. With the passage of time, the application scale of greenhouse vegetable seedling cultivation technology has been expanding. This advanced seedling technology has become an important part of the modern greenhouse planting system. In the greenhouse, people are through the hole tray to grow seedlings, compared with the traditional seedling, the hole tray seedling covers a small area, easy to mechanized unified management. When the seedling grows to a certain period, it should be transplanted. At present, artificial transplanting is the most important and most common transplanting method in greenhouse, but there are many challenges in artificial transplanting, among which the labor intensity is high, the cost is high, the operation efficiency is low and the planting quality is difficult to guarantee is an urgent need to solve the problem. With the vigorous development of facility agriculture, facility hole tray seedling transplanter has become an inevitable trend in the future development of our facilities.

The technology of facility transplanting machine is relatively mature in foreign countries. It not only has advanced computer technology and industrial technology support to make transplanting efficiency high, but also can detect the quality of pot seedlings and record the number of transplanting when transplanting. Relatively speaking, our country started late in the development of facility transplanter, and the technology of transplanter is relatively backward. In spite of this, Chinese researchers are also actively learning from foreign countries in the development of facility transplanter, and constantly conducting experiments and research. Because of the different growth and planting methods of different crops, the development of facility transplanter should be considered a lot, which leads to the slow development of facility transplanter in our country. The research of facility transplanter in China was carried out relatively late, and it was not until the early 1990s that the research of facility transplanter was carried out.

Yunxiang Fan and other researchers of Jilin Agricultural University developed a fully automatic transplanter in 1996. The transplanter adopts a vacuum seedling casting mechanism, which can realize the transplanting of seedlings in the seedling tray to the planting place. Due to the vacuum seedling casting mechanism, the fully automatic transplanter in the greenhouse has a very low seedling injury rate and seedling leakage rate.

Lihui Qiang et al. from China Agricultural University studied the vegetable seedling transfer machine, and invented and designed an automatic lettuce seedling transfer machine in 2005. The seedling transfer machine realized automatic seedling retrieval mainly through two seedling tray conveying mechanisms and one seedling transfer mechanism, as shown in Fig. 5.5. When it works, the seedling tray is put on the conveyor belt and then the switch is pressed to start the work. The sensor is used for induction positioning, and the seedling pulling position is stopped, and then the mechanical arm is used for operation to complete the transplanting. The transplanting efficiency is 3–4 times higher than that of manual transplanting.

Scholars from Nanjing Agricultural University also carried out research on transplanter. In 2009, Ting Zhou, Xiaochan Wang et al. designed a kind of transplanter for pot-tray seedlings, as shown in Fig. 5.6. The horizontal stepper motor controls the screw rotation to realize the horizontal movement of the grasping device, the lifting stepper motor controls the up and down movement of the grasping device, and finally the mechanical claw realizes the action of taking seedlings.

In 2012, Qingchun Feng et al. designed an automatic transplanting machine for flower seedlings, as shown in Fig. 5.7. The transplanter can identify seedlings and holes through machine vision to provide data for transplanting positioning. At the same time, it adopts flexible claw to grasp pot seedlings and reduce the damage to pot seedlings in the transplanting process. Finally, the recognition rate of qualified seedlings by the transplanter reaches more than 87%.

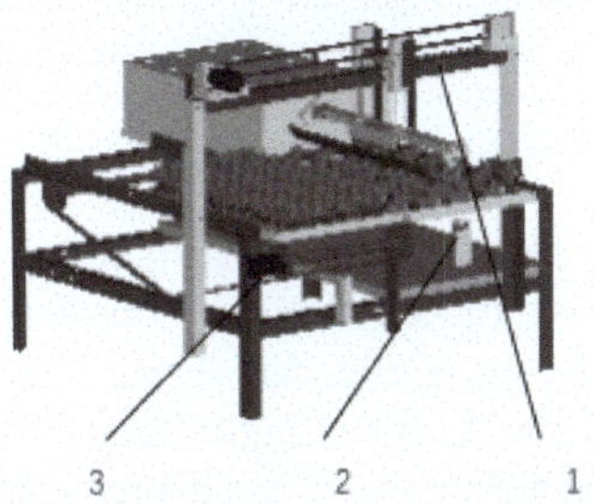

1. Seedling transplanting organization 2. Transmission of the seedling tray
3. Dense seedling tray conveying mechanism

Fig. 5.5 Lettuce automatic seedling transplanter

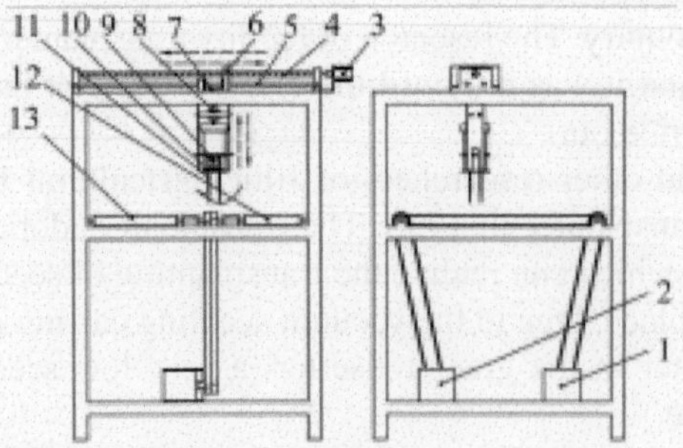

1、2. Conveyor belt motor 3. Horizontal stepping motor 4. Screw transmission mechanism

Guide rail 6. U-shaped slider 7. Lifting stepper motor 8. Screw-nut mechanism 9.

motor 10. Power disk 11. Finger 12、13. Conveyer belt

Fig. 5.6 The whole machine diagram of plug seedling transplanting machine

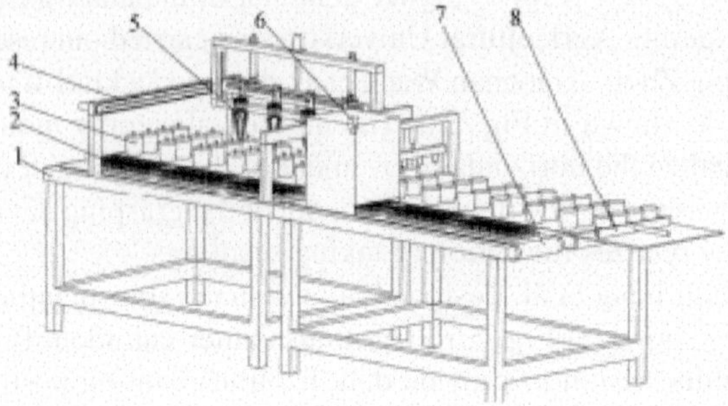

1. Support frame 2. Tray 3. Flowerpot 4. Transplanting positioning mechanism

5. Clamping claw 6. Seedling recognition camera 7. Conveyer belt

8. Feeding push rod

Fig. 5.7 Flower seedling automatic transplanting machine

In 2019, Zhou Xin from Jiangsu University designed a greenhouse pan transplanter, which is a comprehensive device that can complete sorting, transplanting and replanting operations, as shown in Fig. 5.8. The positioning error test shows that the positioning accuracy of the device is high. The results of operation performance test show that the best performance of the device can be achieved when the frequency of transplanting cycle is 1000 times per hour, and the seedling formation rate of 99.33% after replanting can meet the requirements of transplanting operation.

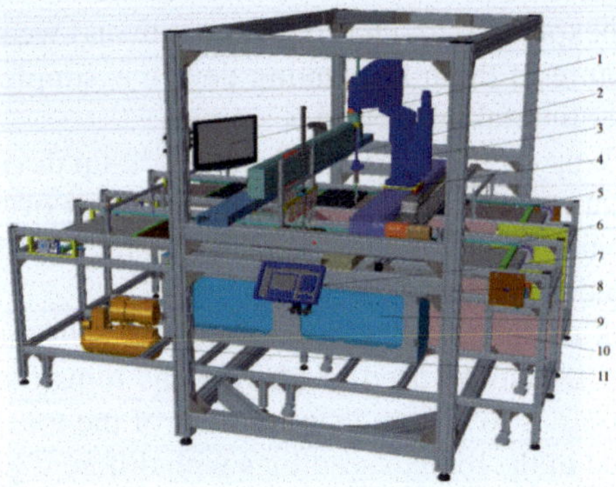

1. display screen 2. Multi-claw transplanting mechanism 3. Single claw replanting mechanism 4. detection device 5. Target tray conveying mechanism 6. Source tray conveying mechanism 7. Touch screen 8. Bench scale 9. Control module 10. Abandoned seedling collection device 11. Air system

Fig. 5.8 Intelligent sorting-transplanting-replanting integrated machine for greenhouse plug seedlings

5.3 Structure Design of Seedling Pick-Up Device

The automatic seedling transplanting machine is composed of a seedling taking device, a seedling throwing device and a planting device. The seedling taking device is the core part of the whole transplanting machine and the basis of transplanting. The whole structure, circuit and control part of the seedling taking device are designed.

5.3.1 The Overall Scheme Design of Seedling Picking Device

1. Design requirements of seedling picking device

In view of the problem that most of the seedlings need to be taken manually by transplanter in existing facilities, a transplanter taking device was designed to replace the manual taking by mechanical device, further reduce the labor cost and realize the automatic taking of seedlings. The seedling taking device uses a manipulator instead of a hand to take seedlings. It takes out the tomato tray clip that reaches the seedling taking position and sends it to the seedling casting position to complete the seedling taking action of the transplanter. At present, for the design of the seedling taking mechanism of the facility transplanter, the following points are mainly considered:

(1) The material is inexpensive and has high strength and wear resistance;
(2) The overall structure of the transplanting device is simple, and the movement process of the seedling taking device is simple;
(3) The seedling pick-up method of the seedling pick-up device is clamping stem type. When picking up seedlings, the clamping effect of the mechanical claw on the stem of tomato seedlings is considered to reduce the damage to the pot seedlings when picking up seedlings;
(4) The seedling pick-up device should be combined with the depth camera, and the depth camera should be used to identify the tomato seedlings before the seedling pick-up, so as to ensure that the stem of the tomato seedlings can be accurately grasped in the tomato seedlings with different growth conditions, so as to avoid damaging the tomato seedlings;
(5) Optimize the design of the seedling pick-up device to ensure that each mechanism does not interfere with each other during operation to ensure the stability of the transplanting process.

2. The overall structure of the seedling device

The function of the seedling taking device is to remove the pot seedling from the hole tray during the transplanting process, and accurately send the pot seedling to the seeding position. The designed seedling picking mechanism adopts the way of holding the pot seedling stem to complete this action. According to the design requirements, a seedling picking device is designed, and the overall structure of the seedling picking device is shown in Fig. 5.9.

The overall structure of the seedling picking device should not only consider the selection of each part, but also consider the simplicity of the device. It can be seen from Fig. 5.9 that the seedling picking device is composed of a frame, a conveyor

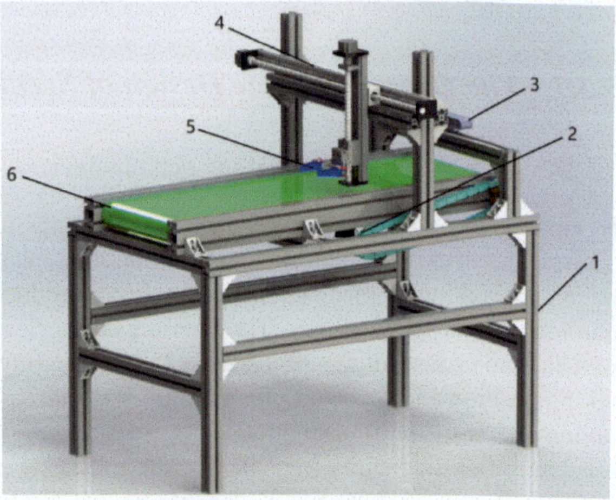

Fig. 5.9 The overall structure diagram of the seedling pick-up device

1. Frame 2. Stepper motor 3. Camera
4. Ball screw 5. Flexible jaw 6. Conveyer belt

belt, a stepper motor, a ball screw, a flexible gripper, and a camera. The conveyor belt is fixed on the frame, and is driven by the motor to complete the conveying of the pot plate seedlings, and the pot plate seedlings are transported to the position of taking the seedlings. The stem of the tomato seedling is clamped at the position of the seedling, and the automatic seedling is completed. The identification was completed by the camera behind the frame before picking up the seedlings. Through the cooperation of various components, the automatic seedling picking process of tomato plug seedlings is realized.

3. The working principle of seedling picking device

The working process of the seedling device is as follows: (1) Place the hole plate seedling on the conveyor belt, and limit the hole plate surface by the baffle fixed on the conveyor belt. The installation diagram of the hole plate is shown in Fig. 5.10. (2) The camera fixed at the rear of the frame begins to identify and position the hole plate seedling. (3) The ball screw drives the mechanical claw to move horizontally and vertically to the corresponding position, and the conveyor belt starts to move to stop when the limit switch is triggered, and the mechanical claw is just able to grab the tomato stalk. (4) The mechanical claw closure begins to return to the starting point along the longitudinal ball screw, and then returns to the origin along the transverse ball screw at the same time the conveyor belt returns to a hole distance, at which time the process of taking a seedling ends.

When a seedling is taken, the previous action is repeated until a row of seedlings (6 seedlings) is taken. When the second row of seedlings is taken, the first step is to identify and locate, and then the mechanical claw reaches the corresponding position through horizontal and vertical movement. After the baffle on the conveyor belt triggers the limit switch, the distance of a hole (38 mm) is moved forward, so that the mechanical claw can just grasp the stem of tomato seedlings. The next step is to complete the picking of seedlings and return to the origin. At the same time, the distance of the conveyor belt back to a hole is repeated in this row. When the current operation line is finished, the next line of seedling picking operation is carried out. The same as the seedling picking action, the change is only the forward distance after the conveyor belt triggers the limit switch. The forward distance after the trigger limit switch can be regarded as the distance of n-1 holes n is the number of rows of current seedling picking, and the distance of each return is the distance of 1 hole. The control structure diagram of the seedling picking device is shown in Fig. 5.11.

5.3.2 *Selection of Core Components of Seedling Pick-Up Device*

Whether the seedling pick-up device can complete the seedling pick-up work depends on each key component. In the design and selection of the transplanting machine, the selection of key components is very important, which directly affects the stability,

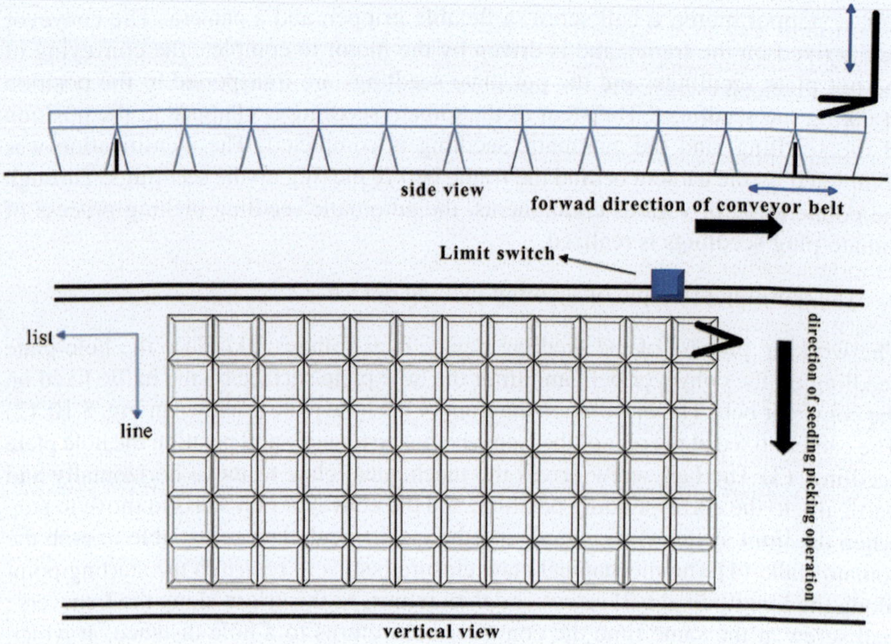

Fig. 5.10 Plug tray installation diagram

accuracy and efficiency of the transplanting machine. According to the design requirements, the key components are selected.

1. The selection of controller

In order to realize the intelligent control of the seedling picking device, the controller is usually used for control. The commonly used controllers are Arduino, Raspberry Pi, STM 32 and other embedded development boards, or professional industrial controllers and PLC. In order to meet the design requirements and realize the precise control of the stepper motor, the controller should choose a simple programming environment, which is convenient for the operator to write and debug the control program, so as to realize the precise control of the seedling picking mechanism. Therefore, the controller selects the Arduino Uno development board, the model is Arduino Uno R3, as shown in Fig. 5.12, and its parameters are shown in Table 5.1.

2. Selection of Ball Screw

The seedling taking device adopts a seedling taking mechanism composed of gantry frame structure. The seedling taking mechanism takes the flexible gripper to reach the seedling taking position by moving the ball screw. Figure 5.13 shows the composition of the seedling taking mechanism.

As can be seen from Fig. 5.14, the selection of ball screw travel should take into account the size of the rack and the size of the slider. As shown in Fig. 5.14a, the

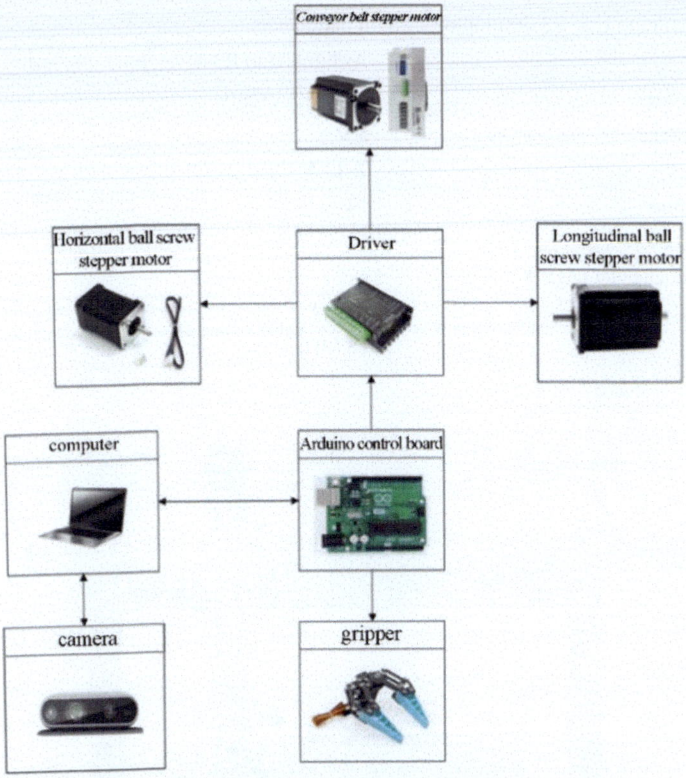

Fig. 5.11 Control structure diagram of seedling pick-up device

Fig. 5.12 Arduino Uno R3 development board

Table 5.1 Arduino Uno R3 development board parameters

Controller	Arduino Uno R3 development board
Microcontroller	ATMEGA328P
Working voltage	5 V
Input voltage	7–12 V
Digital I/O pins	14
PWM channel	6
Analog input channel (ADC)	6
Each I/O DC output capability	20 mA
Clock speed	16 MHz

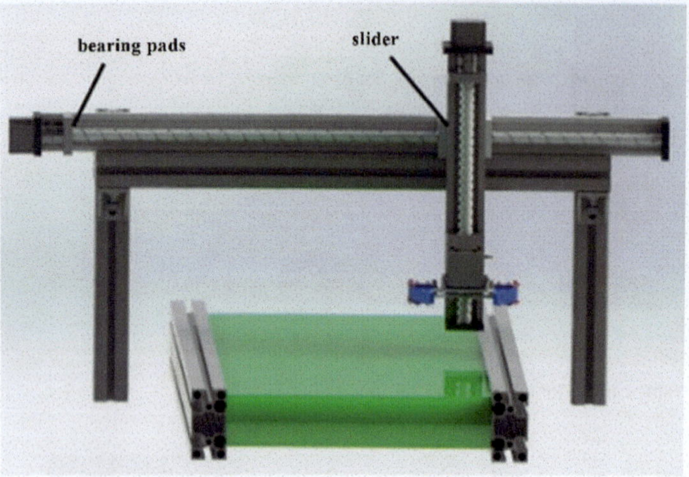

Fig. 5.13 Grab seedlings institutions

width of the frame is 580 mm, the size of the conveyor belt is 420 mm, and the size of the slide block is 70 mm. At the same time, considering that the mechanical claw should return to the origin for seeding and the installation position of the conveyor belt motor and chain after taking the seedling, the travel of the transverse ball screw should be greater than 650 mm, and the buffer of 25 mm is left and right. After comprehensive consideration, the travel of the ball screw for lateral movement is selected as 700 mm.

It can be seen from Fig. 5.14 that the longitudinal ball screw is mainly used for the rise and fall of the mechanical claw. Considering the influence of plug height and seedling height in the stroke selection of longitudinal ball screw, it is necessary to avoid the end of the ball screw contacting the conveyor belt, and at the same time, the mechanical claw can grasp the stem of tomato seedlings. Therefore, combined

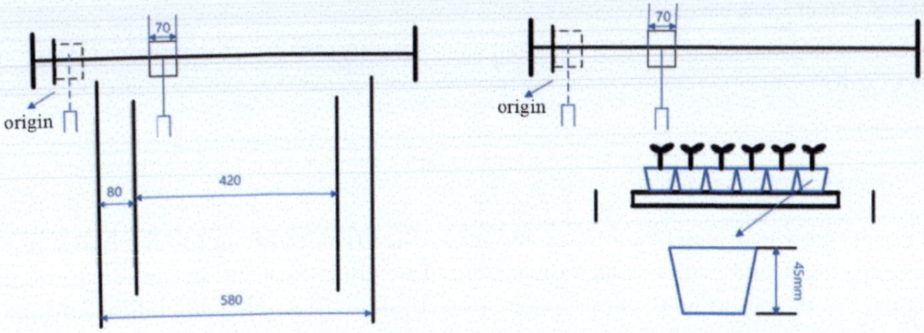

a. Overlook

b. Envisaging

Fig. 5.14 Picking seedling mechanism diagram

with the height of the gantry and the requirements of picking seedlings, the stroke of the longitudinal ball screw is selected as 400 mm.

The ball screw model CBX1610 is selected, as shown in Fig. 5.15, and the parameters are shown in Table 5.2.

Fig. 5.15 Ball screw

Table 5.2 Ball screw parameters

Model	Maximum load (KG)	Ball screw accuracy (mm)	Linear guide (MM)
CBX 1610	50	0.03	12

3. Stepper motor selection

The ball screw is driven by a motor, and under the drive of the motor, the ball screw can move back and forth to reach the specified seedling position. In the selection of stepper motor, to choose the appropriate step Angle, step Angle calculation formula is as follows:

$$\alpha = (360^\circ \times A) \times d \tag{5.1}$$

In the formula (5.1): α is the step angle, °; A is the accuracy of ball screw, mm; d is the lead of the screw, mm.

Substituting the accuracy of the ball screw and the lead of the screw, the step angle is 1.8°.

The output torque of the ball screw stepper motor is an important factor to determine its performance, and the torque under different loads is also different. The torque calculation formula is:

$$T = \frac{F \times d}{2} \tag{5.2}$$

In the formula (5.2): T is torque, N m; F is the load, N.

Substituting the maximum load and lead of the ball screw, the torque can be obtained as 1.75 N m.

Considering the accuracy, torque and transmission efficiency, the 57BYG250C two-phase stepper motor is selected to drive the ball screw, and the DM452 driver is selected. The stepping motor is shown in Fig. 5.16, and the parameters are shown in Table 5.3.

Stepper motor driver is a device used to control the stepper motor. It converts current and pulse signals into signals suitable for stepper motor drive. The main function of the stepper motor driver is to provide appropriate current and control signals for the motor to achieve accurate motion, positioning and control of the stepper motor. The DM 452 driver is matched with the stepper motor. The DM 452 driver is shown in Fig. 5.17, and the parameters are shown in Table 5.4.

4. Camera selection

In order to complete the identification and positioning of the tomato stalk and the grasp of the mechanical claw, the seedling taking device needs the camera as a transmission tool to complete the acquisition of the image of the tomato stalk. Cameras can be divided into three types according to different operating principles, which are monocular cameras, binocular cameras and depth cameras. The monocular camera

Fig. 5.16 Stepper motor

Table 5.3 Stepper motor parameters

Model	Stepping angle (°)	Rated current (A)	Lead number (root)	Phase number (phase)
J-5718HB 2401	1.8	3.5	4	2

can capture the two-dimensional image of the scene and cannot directly obtain the depth information, but it can analyze and process the image through the monocular vision algorithm to obtain the depth information of the object. Monocular cameras are suitable for general object detection, recognition and tracking. The binocular camera is composed of two cameras, simulates the visual system of the human eyes, calculates the parallax information of the object by obtaining the images from two perspectives, and deduces the distance and depth of the object. Binocular cameras are suitable for applications requiring accurate depth information such as stereo vision, depth perception, and 3D reconstruction. Depth camera uses structured light, time flight or stereo vision technology, can accurately obtain the distance information of each pixel, to achieve three-dimensional perception of the scene, suitable for three-dimensional reconstruction, gesture recognition, gesture control and other applications.

In the process of picking the seedling, it is necessary to accurately grasp the stem of the tomato seedling, so it is necessary to obtain the three-dimensional position information of the mechanical claw. To meet this point monocular cameras are obviously not satisfied, so consider binocular cameras and depth cameras. It is found that binocular camera is affected by ambient light, exposure degree and measured scene, which leads to unsatisfactory detection effect. Therefore, the depth camera is selected for the camera of the seedling taking device.

Fig. 5.17 DM452 driver

Table 5.4 DM452 driver parameters

Model	Voltage (°)	Current (A)	Division number
DM452	18–48	1.0–4.2	1–256

Since the transplanting operation range of tomato pot seedlings is small, usually between 0.3 and 1.0 m, the close-positioning camera can meet the operation needs. At the same time, considering the requirements of precision and device cost in the seedling picking operation, Intel RealSense D435i depth camera is selected, as shown in Fig. 5.18. Table 5.5 lists the parameters.

The Intel RealSense D435i depth camera consists of an infrared camera, a dot matrix projector, and an RGB camera. The Intel RealSense D435i depth camera uses the dot matrix projector and left and right infrared camera to obtain the corresponding image by combining the vision processor and color sensor under the calculation of triangulation. Figure 5.19 shows the principle of triangulation. Where: P is a point

Fig. 5.18 Intel RealSense D435i depth camera

Table 5.5 Intel RealSense D435i parameters

Name	D435i
Depth camera	1920 × 1080
RGB maximum resolution	
Depth resolution	1280 × 720
Depth of field rate	90 FPS
Depth detection range	0.2–10 m
Connector	USB 3.1 Gen 1

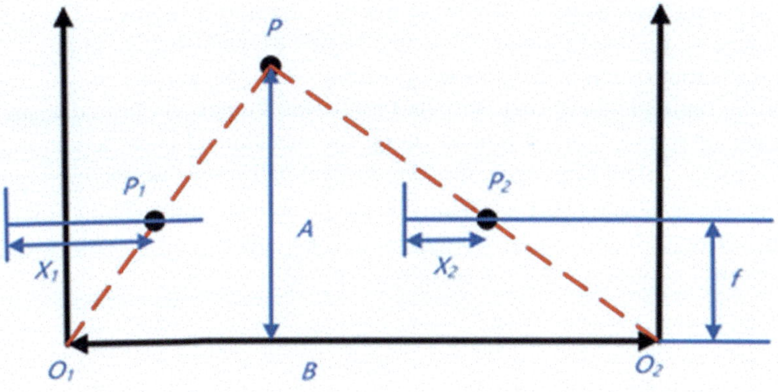

Fig. 5.19 Triangulation

on the target object, O_1 and O_2 are the optical centers of the two infrared cameras respectively, P_1 and P_2 are the imaging points of P on the two infrared cameras respectively, X1 and X2 are the horizontal coordinates of the two infrared cameras in the imaging.

f is the focal length of the camera, B is the distance between the two infrared cameras, and A is the required depth value. The procedure for finding the depth value A is as follows: Suppose the distance between P1 and P2 is D. In that way

$$D = B - (X_1 - X_2) \tag{5.3}$$

Through the triangle similarity principle, it can be seen that:

$$\frac{B - (X_1 - X_2)}{B} = \frac{A - f}{A} \tag{5.4}$$

$$A = \frac{f \times B}{X_1 - X_2} \tag{5.5}$$

5.4 Design of Control System for Seedling Picking Device

According to the whole structure and working principle of the seedling taking device, the control system of the seedling taking device is designed. In the early stage, the controller was selected, and the Arduino Uno development board was selected to control it. Therefore, the control system with Arduino as the core was built, and the motor and mechanical claw were controlled by Arduino IDE software to complete the action of taking seedling.

The whole process of taking seedlings is: after the camera is identified and positioned, the mechanical claw starts from the origin, and the horizontal ball screw begins to work, driving the mechanical claw to the corresponding position. The mechanical claw opens, the longitudinal ball screw begins to work, begins to move down, reaches the tomato stem in front of the position, the conveyor belt moves forward to touch the limit switch to stop, stop and go forward again n-1 hole distance, n is the number of rows of the current seedling, the mechanical claw begins to close. When the tomato plant is grasped, the longitudinal ball screw moves up to return to the origin, the horizontal ball screw also begins to return to the origin at the same time, the conveyor belt is backward a hole distance, and finally reaches the origin, the mechanical claw is opened for seeding (Fig. 5.20).

5.4.1 *System Circuit Design*

In order to ensure the normal operation of the seedling taking device, it is necessary to ensure the stable and reliable connection between the controller, the driver and the steering gear when the system is started, and at the same time, the controller sends instructions to them to drive the stepper motor to achieve the seedling taking operation. In this device, the Arduino Uno controller provides power to it through 5 V DC power supply, and the stepper motor power supply on the two ball screws uses 24 V DC power supply. The 220 V AC in the socket is converted to 24 V DC through the power switch to supply power to the stepper motor and limit sensor. The PUL-and DIR-on the motor driver are connected to each other and connected to the GND pin of the Arduino Uno controller. The PUL+ and DIR+ on the lateral ball screw stepper

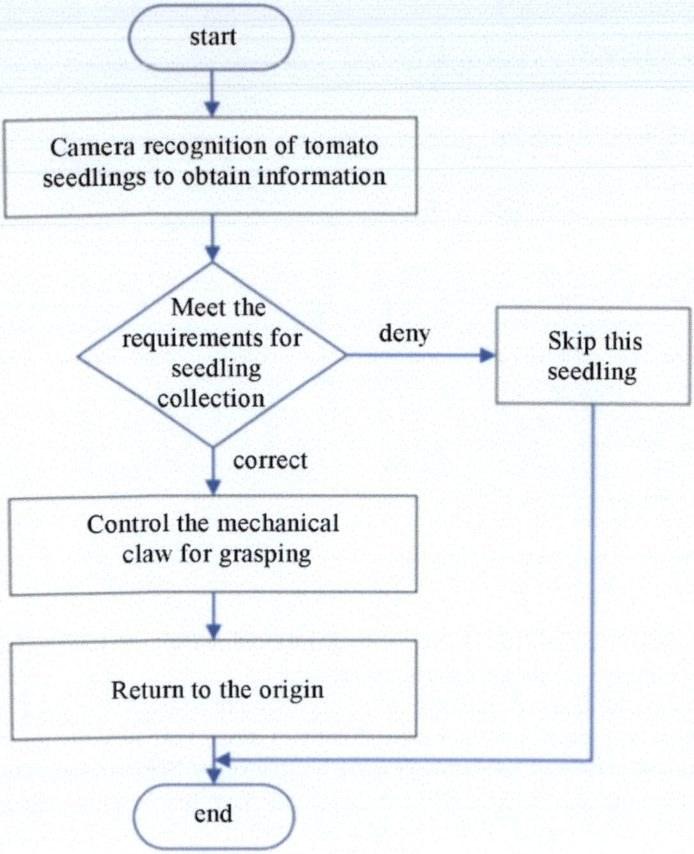

Fig. 5.20 The working flow chart of seedling picking device

motor driver are connected to the pin 9 and pin 8 of the Arduino Uno controller, respectively. The PUL+ and DIR+ on the longitudinal ball screw motor driver are connected to the pin 11 and pin 10 of the Arduino Uno controller, respectively. A+, A−, B+ and B− are respectively connected to the four terminals corresponding to the stepper motor. The mechanical claw is opened and closed through the steering engine. Its signal cable is connected to pin 7 of the ArduinoUno controller, and the power cable and ground cable are connected to the 5 V pin and GND pin of the ArduinoUno controller respectively. The limit sensor of the horizontal ball screw is connected to pins 2 and 4 of the ArduinoUno controller, and the limit sensor of the longitudinal ball screw is connected to pins 3 and 5 of the ArduinoUno controller. Figure 5.21 shows the principle of the control circuit.

If the whole seedling taking device wants to operate normally, it needs a controller to control each component. By controlling the output pulse signals of the Arduino

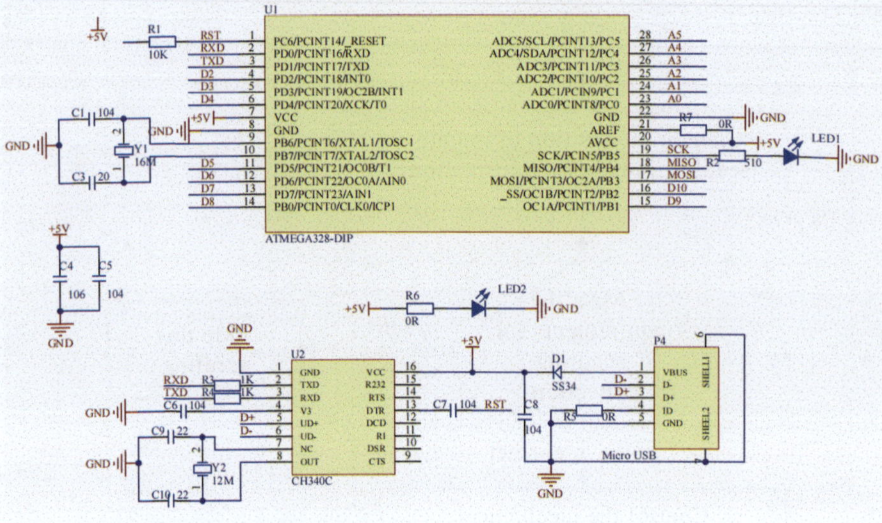

Fig. 5.21 Control circuit schematic diagram

Uno development board, the stepper motor, infrared sensor, and steering gear are controlled. Figure 5.22 shows the controller ports.

The device adopts three stepping motors, which are respectively used to control the transverse ball screw and the longitudinal ball screw to realize the movement of the mechanical claw and the conveying of the conveyor belt; As the actuator that drives the opening and closing of the mechanical claw, the steering gear is usually controlled by PWM signal. This control method realizes the precise control of the position and Angle of the steering gear by adjusting the pulse width of the input signal. Figure 5.23 shows the wiring diagram of the stepper motor and steering gear.

In order to get the signal that the device moves in place and avoid the accident caused by the mechanism moving out of the range, the seedling taking device uses an infrared sensor to limit the position of the ball screw. Figure 5.24 shows the connection of the infrared sensor.

5.4.2 Control Software

Arduino IDE is a software tool designed for programming Arduino microcontrollers. It provides a clean, comprehensive interface that makes it easy for users to edit, compile, upload, and debug code, helping developers develop and deploy Arduino projects quickly and efficiently. Arduino usually uses the C/C++ language for software programming. C/C++ is a general-purpose programming language with a wide range of applicability, and embedded systems and low-level programming are often adopted in this language. With C/C++, developers can take full advantage

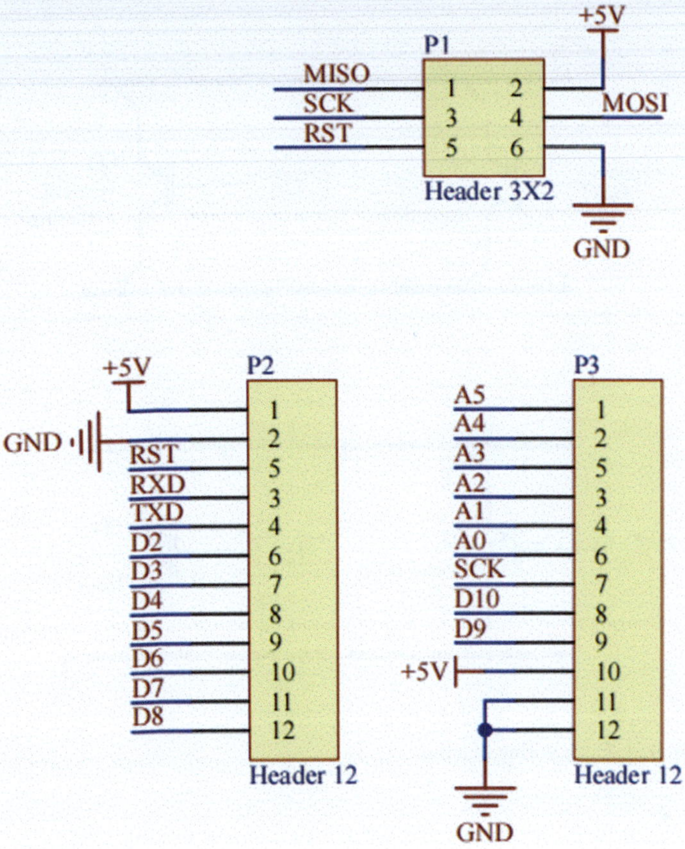

Fig. 5.22 Controller interface diagram

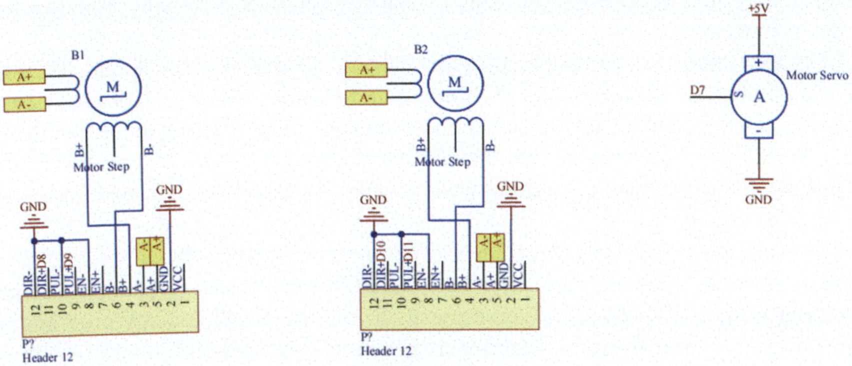

Fig. 5.23 Stepper motor and steering gear wiring diagram

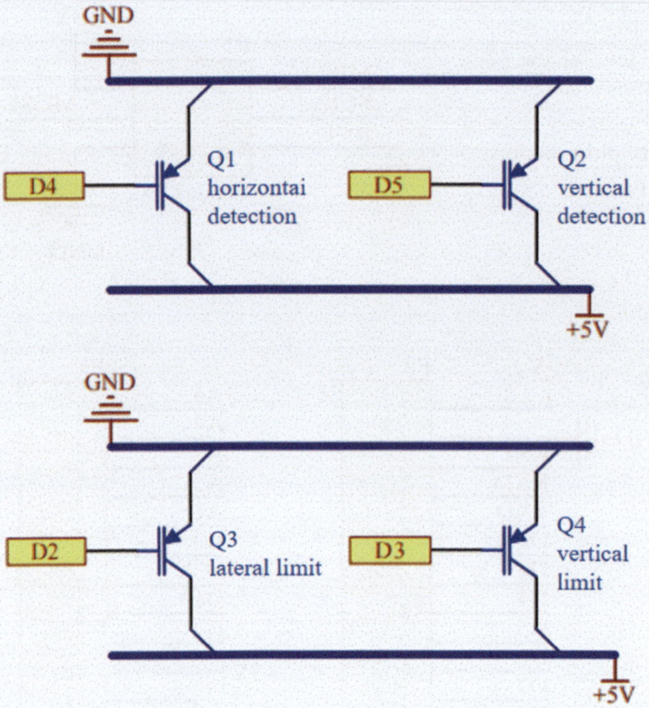

Fig. 5.24 Infrared sensor wiring diagram

of Arduino's hardware resources and write efficient and flexible code. The Arduino IDE interface consists of a toolbar, a code editor, and a message notification area. The Arduino program structure includes the setup () function, which is used to initialize Settings, and the loop () function, which is used to loop the input and output of the program functions.

ROS, as an open-source system, provides a flexible, versatile and scalable platform for robot software development. ROS has four functions: communication mechanism, development tool, application function and ecosystem. Among these four functions, ROS's point-to-point communication mechanism is one of its core parts, and point-to-point communication can be easily achieved through the Publish/Subscribe communication model.

The communication between ROS and Arduino is based on the communication mode of ROS, and the communication between ROS and Arduino is realized through Rosserial library. The Rosserial server needs to run the serial_node.py program of the rosserial_pyhton function package to generate corresponding nodes and then communicate with the device running the rosserial client. The Rosserial client uses rosserial_arduino to make ROS work with the Arduino IDE. Rosserial_arduino integrates the stepper motor control method into the control system of ROS environment by making Arduino UNO in the form of subscription topics. By receiving the information of each node, ROS can detect and recognize the pixel position of the tomato stem on the RGB visible image and extract the spatial coordinates of the tomato stem in the three-dimensional coordinate system with the depth camera as the origin. Finally, the spatial coordinates of the tomato stem are published. After receiving the information, the next node reads the modeling data to realize the position relationship between the depth camera and the tomato stem, converts the space coordinates to get the true distance between the end actuator and the tomato stem, and publishes it in the form of a question. Finally, according to the position of the end actuator to the stem of the tomato plant, the control command is issued to the mechanical claw to realize the action of taking the plant. The seedling taking device is energized, and the seedling taking device is reset, so that each component of the device returns to the initial position. The initial seedling taking position is determined by the infrared sensor, and the seedling taking work begins after the device is reset.

5.5 Verification Experiment of Seedling Picking Device

The above completed the selection of parts of tomato plug seedling picking device, the design of control system, the research of deep learning algorithm and tomato seedling target positioning algorithm. The next step is to build a seedling device test bench and verify the performance of the seedling device. The rationality of the design scheme is verified by the results of the seedling test, and the possible shortcomings of the seedling device are found and corrected according to the test results.

5.5.1 Construction of Test Bench for Seedling Pick-Up Device

According to the previous design, hardware selection, circuit design and program design, the whole seedling picking device is built, the whole control system is improved and built, and each hardware part is debugged. Finally, it can meet the requirements of the previous design. The overall physical diagram of the seedling picking device is shown in Fig. 5.25, and the frame size is $1000 \times 580 \times 540$ mm (length $\times$ width $\times$ height).

When the test frame was built, in order to meet the dimensional accuracy, the frame was built with European standard aluminum profile 4040, processed and manufactured by laser cutting to ensure the accuracy of its size, and assembled by T-screws. Combined with the size of the tray and the size of the conveyor belt, the size of the tray baffle is finally determined to be 280 mm long and 20 mm high.

Among them, the conveyor belt is fixed in the middle of the frame, the baffle plate for fixing the hole plate is fixed above the conveyor belt, the motor of the conveyor belt is located below the conveyor belt, the ball screw slide is installed above the frame, the two stepper motors are installed on the two ball screws, and

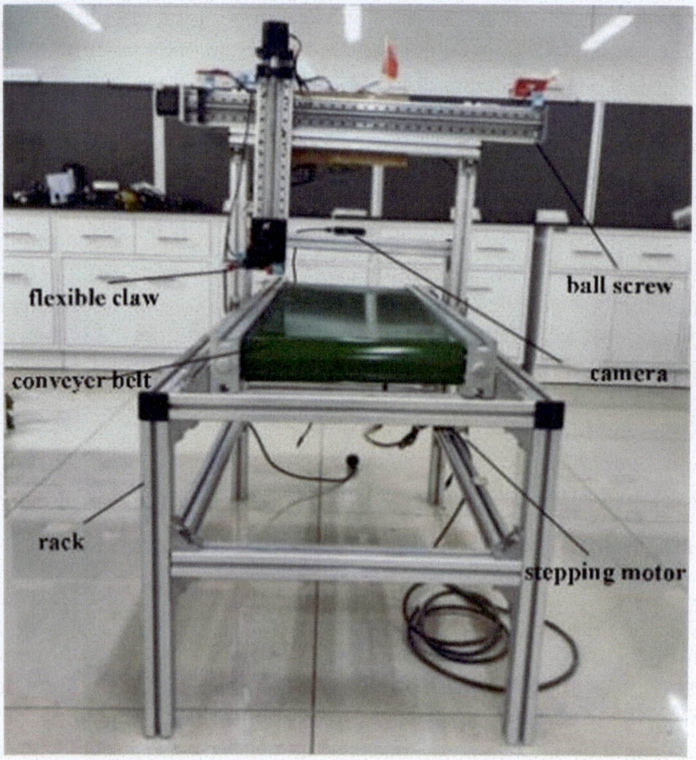

Fig. 5.25 The overall physical map of the seedling picking device

the longitudinal ball screw is fixed on the nut support of the transverse ball screw. The mechanical claw is fixed on the nut support of the longitudinal ball screw, and is driven by the nut to complete the work of the mechanical claw. The establishment of the test bench of the device for taking seedlings provided the basis for the device for taking seedlings.

5.5.2 *Experimental Study on Performance of Seedling Pick-Up Device*

The seedling picking test was carried out for the built seedling picking device, and the performance of the control system of the whole seedling picking device was verified. The test results are an important basis for checking whether the seedling picking device meets the design requirements.

5.5.3 Experiment Purpose

The seedling picking performance of the seedling picking device and the damage of the pot seedling are very important to the transplanting machine. There are three purposes:

(1) Check whether the overall structure of the seedling taking device is reasonable and reliable;
(2) The identification and positioning of tomato seedling stems were tested by taking seedling tests;
(3) Check whether the device can achieve seedling taking through seedling taking test, and check the damage to tomato seedlings during seedling taking process.

5.5.4 Evaluating Indicator

(1) The success rate of taking seedlings: take the tomato seedlings out of the hole dish and send them to the designated position, and complete the whole process as taking seedlings successfully. The calculation formula is as follows:

$$\alpha = \frac{n}{N} \times 100\% \tag{5.6}$$

In the formula (5.6): α is the success rate of taking seedlings, %. n is the number of tomato seedlings successfully taken. N is the total number of tomato seedlings.

(2) Damage rate of tomato seedlings: In the process of picking seedlings, the damage of tomato seedlings will occur when the mechanical claw grabs the tomato seedlings, and the damaged tomato seedlings will affect the subsequent planting. Therefore, the damage rate of tomato seedlings is used as the evaluation index of the experiment. The damage of tomato seedlings is defined as: the stem of tomato seedlings breaks and the leaves are damaged. The calculation formula of damage rate is as follows:

$$\beta = \frac{m}{M} \times 100\% \tag{5.7}$$

In the formula (5.7): β is the damage rate of tomato seedlings, %; m is the number of tomato seedlings damaged in the process of taking seedlings; m is the total number of tomato seedlings.

5.5.5 Experiment Scheme

On the test platform, the seedling taking experiment of tomato cavity plate was carried out. According to the designed program, the seedling taking test was started, the pot plate seedling was placed on the seedling taking device, the stem information of the tomato seedling was obtained by the camera, and then the mechanical claw was controlled to grasp the tomato seedling and observe the seedling taking situation of the device. In this experiment, the selected seedling age was 40 days.

(1) Identification of tomato stems. In the detection system, the tomato stalk in the image is obtained, and the frame selection mark position is carried out. The detection accuracy of tomato stems was calculated.
(2) Seedling taking device. The seedling taking device takes seedlings according to the designed seedling taking process. According to the seedling taking condition of the seedling taking device, the success rate of seedling taking and the damage rate of seedling taking are used as evaluation indexes to analyze the results.

5.5.6 *Experiment Results and Analysis*

(1) Identification of tomato seedling stems

The tomato hole pan seedling was placed on the seedling taking device, and the stem of the tomato hole pan seedling was identified and positioned by the depth camera on the seedling taking device. In this experiment, a 6 × 12 tomato tray seedling was used as the test object, and 10 groups were carried out in total. In each experiment, the depth camera was used to identify and position the stem of the tomato seedling in the outermost row. When the camera identified and photographed, the position of the tray was at the position where the tray normally arrived when the device took seedlings, and each tray was defined as seedling 1~ seedling 6 from left to right. Finally, their confidence and detection time were recorded (Table 5.6).

It can be seen from the table that the recognition accuracy of the device to the tomato stem needs to be improved, but the detection time is short, and the average detection time is 1.06 s. It can be seen from Fig. 5.26 that the tomato seedling stem can be well identified, and the overall recognition effect is good. Through analysis, it is found that the reason for the influence of confidence may be due to the influence of the seedling environment. Too light or too dark will affect the recognition of tomato seedling stems, resulting in low recognition accuracy.

(2) The situation of seedling pick-up by the device

The tomato pot-plate seedling was placed on the designed seedling taking device, and the seedling taking test was carried out according to the set working mode. The experiment used 6 × 12 tomato pot plate seedlings as the test object. There were 10 groups of seedling extraction test, and the water content of matrix of each group was the same. The seedlings were taken by running the seedling taking device, observing the seedling taking process and calculating the evaluation indicators, and the results were recorded in Table 5.7 (Fig. 5.27).

Table 5.6 Identification test of tomato seedling stem

Serial number	Confidence						Detection time (s)
	Seedling 1	Seedling 2	Seedling 3	Seedling 4	Seedling 5	Seedling 6	
1	0.55	0.58	0.56	0.59	0.65	0.64	1.03
2	0.61	0.62	0.51	0.57	0.57	0.62	1.12
3	0.56	0.65	0.54	0.53	0.59	0.65	1.08
4	0.63	0.67	0.63	0.62	0.57	0.61	0.98
5	0.59	0.61	0.64	0.57	0.61	0.58	1.05
6	0.64	0.56	0.65	0.63	0.59	0.62	1.14
7	0.58	0.63	0.62	0.59	0.64	0.58	0.96
8	0.55	0.65	0.64	0.57	0.59	0.59	1.03
9	0.61	0.64	0.55	0.61	0.57	0.60	1.10
10	0.84	0.58	0.62	0.65	0.59	0.51	1.07
Average							1.06

Fig. 5.26 Seedling identification

Table 5.7 Mechanical claw seedling results

Test times	Picking success rate (%)	Damage rate (%)
1	94.4	8.8
2	90.3	6.1
3	93.1	7.4
4	94.4	7.3
5	90.3	9.2
6	95.8	5.8
7	94.4	5.9
8	91.7	7.6
9	93.1	6.0
10	91.7	7.6
Average	92.9	7.2

a. Reaching in front of the tomato seedlings

b. Reaching the picking seedling position

c. Clamping tomato seedlings

d. Seedling return journey

e. Reaching the seedling position

Fig. 5.27 Seedling picking process

It can be seen from the table that the damage rate of the seedling picking device is 7.2%. At the same time, the success rate of the device was between 90 and 96%, and the average success rate was 92.9%. Through the analysis, it was found that the factor affecting the success rate of taking seedlings was the system's identification results of the tomato seedling stalk. When the system did not identify the tomato seedling with poor growth condition, the mechanical claw would not accept the command to take out the seedling with poor growth condition, and then it would not affect the success rate of taking seedlings because it could not grasp the seedling with poor growth condition. Therefore, more in-depth research on tomato stalk recognition is needed.

5.6 Research Conclusion of Intelligent Transplantation Device for Desert Gobi Cave Seedlings

Aiming at the needs of facility agriculture in southern Xinjiang, this study designed and verified an intelligent transplanting device for desert gobi cave seedlings. The device uses Arduino to control the work of each component and combines the depth camera to realize the picking of tomato hole plate seedlings. According to the current situation of facility tomato transplanting, the design of plant taking device of facility tomato transplanting machine was carried out. The design requirements of the device for taking seedlings were put forward. The overall structure of the device is designed. The device is mainly composed of frame, stepper motor, camera, ball screw, flexible clamp and conveyor belt. According to the control requirements and design requirements of the overall structure of the device, the core components of the device were selected, including controller, ball screw, stepping motor, mechanical claw and camera. The control system of the seedling taking device was designed, and the control system was designed. The whole seedling taking device was integrated based on ROS robot system. The circuit design of the control system is completed by analyzing the control requirements and logic, and the program code of the control system is compiled and tested by Arduino. The design of the seedling taking device fully considers the stability and operation accuracy of the equipment, and optimizes the cooperative work between the manipulator and the conveyor belt to reduce the damage of seedlings and improve the efficiency of seedling taking.

The experiment verified the performance of the device, and the results showed that the device showed high accuracy in identifying and grasping the stems of tomato seedlings, the average success rate of picking seedlings reached 92.9%, and the damage rate was 7.2%. These results show that although the device has shown some effectiveness in practical applications, there is still room for improvement in further improving the recognition accuracy and reducing damage. The study also found that the light conditions had a significant impact on the identification accuracy of tomato stems, so future work could be further optimized to improve the adaptability of the system in different environments. This study provides a certain technical reference for the intelligent development of facility agriculture in southern Xinjiang, but in terms of the promotion and application of technology and further optimization, further research is still needed.

Chapter 6
Study on the Water and Fertilizer Demand Model for Facility Fruit and Vegetables in Desert and Gobi Regions

Wenhao Xia, Can Hu, Ji Shi, Shufeng Zhuang, Xiaoqin Lian, and Runrun Wang

Facility agriculture is a pivotal technology in modern farming, enabling year-round production through greenhouse and other facility-based cultivation. This approach is particularly beneficial in regions with unstable or extreme climates, such as the cold northern areas and the arid southern Xinjiang of China. Facility agriculture allows for precise control over temperature, humidity, and light, creating optimal conditions for fruit and vegetable growth. This results in significant improvements in yield and quality while reducing pest and disease occurrences. With effective management and optimized cultivation techniques, facility-grown produce can achieve high yields and efficiency, meeting market demands for high-quality fresh produce, especially during winter and early spring when fresh produce is scarce. Additionally, facility agriculture often incorporates advanced methods like soilless cultivation, drip irrigation, and integrated water and fertilizer systems, leading to more efficient use of water and

W. Xia · C. Hu (✉) · J. Shi · S. Zhuang · X. Lian
College of Mechanical and Electrical Engineering, Tarim University, Alar 843300, China
e-mail: hucanboy1@qq.com

Modern Agricultural Engineering Key Laboratory at Universities of Education Department of Xinjiang Uygur Autonomous Region, Alar 843300, China

W. Xia · J. Shi · S. Zhuang · X. Lian
Xinjiang Production and Construction Corps (XPCC) Key Laboratory of Utilization and Equipment of Special Agricultural and Forestry Products in Southern Xinjiang, Alar 843300, China

R. Wang
Key Laboratory of Smart Agriculture System Integration, Ministry of Education, Beijing 100083, China

Key Laboratory of Agricultural Information Acquisition Technology, Ministry of Agriculture and Rural Affairs, Beijing 100083, China

C. Hu et al. (eds.), *New Technologies Applied in Gobi Desert Facility Agriculture*, Smart Agriculture 11, https://doi.org/10.1007/978-981-96-1031-0_6

fertilizers, reduced environmental impact, and lower production costs. The ability to supply the market year-round, particularly during off-seasons, provides significant economic benefits and has become a crucial income source for many farmers and agricultural enterprises.

This chapter will use facility-grown strawberries as an example. Strawberries are a high-value crop with significant market demand and a relatively short growth cycle, typically taking only a few months from planting to harvest. This rapid growth cycle makes strawberries an efficient crop for experimental purposes in facility agriculture, allowing for quick observation of how facility conditions affect growth and yield. This provides timely feedback for improving facility agriculture technologies. Strawberry cultivation in facilities can yield substantial economic benefits, so using facility-grown strawberries as an example effectively demonstrates the impact of facility agriculture technologies and supports the development and application of these methods. In modern agricultural production, water and fertilizer management is crucial. Both water and fertilizers are fundamental to plant growth, and their effective management directly impacts crop growth, yield, and quality. It also enhances resource use efficiency and reduces environmental pollution. Scientific water and fertilizer management ensures that crops receive the right amounts of water and nutrients at different growth stages, promoting healthy growth, improving photosynthetic efficiency, increasing yield, and enhancing fruit quality. Precision irrigation technology adjusts water application based on crop needs, avoiding over-irrigation and conserving water resources. Proper fertilization improves nutrient use efficiency, reduces runoff, and minimizes soil and water pollution. By optimizing water and fertilizer management, agriculture can become more sustainable, reducing excessive fertilizer and water use, lowering production costs, and mitigating negative environmental impacts such as water body eutrophication and soil salinization. With global climate change, agricultural production faces increasingly complex challenges. Water and fertilizer management technologies help farmers cope with extreme weather conditions such as droughts and floods, ensuring stable crop production and food security. Modern agriculture is moving towards smarter and more precise approaches, with the application of new technologies like integrated water and fertilizer systems, IoT sensors, and intelligent irrigation systems making management more efficient and intelligent. The importance of water and fertilizer management cannot be overlooked. It affects not only crop yield and quality but also agricultural sustainability and environmental protection. In facility-grown strawberry cultivation, proper water and fertilizer management is particularly critical. It meets strawberries' growth needs while maximizing resource utilization, reducing environmental stress, and enhancing economic benefits.

6.1 Background and Challenges of Facility-Grown Strawberry Cultivation

6.1.1 Overview of Facility-Grown Strawberry Cultivation

Facility cultivation allows for precise control of environmental conditions. Temperature, humidity, and light levels within greenhouses and tunnels can be adjusted according to different growth stages of strawberries, avoiding poor growth or pest and disease issues caused by weather changes. This capability significantly increases strawberry yield and fruit quality, enabling high production even under adverse external conditions. Facility-grown strawberries are typically cultivated using soilless techniques. These methods use substrates like coconut coir, perlite, or sand instead of soil, allowing for better control of moisture and nutrient supply to the roots. Soilless cultivation reduces soil-borne diseases and improves fertilizer and water efficiency. Drip irrigation and integrated water-fertilizer technologies are widely used, precisely calculating water and nutrient needs and automatically providing the right amounts. This not only accelerates growth and increases yield but also conserves water and reduces fertilizer costs, minimizing environmental pollution.

From an economic perspective, facility-grown strawberry cultivation has significant market potential. By producing high-quality strawberries during non-traditional growing seasons and meeting market demand, facility-grown strawberries command higher prices and offer substantial economic benefits. The adoption of this model across various regions has also facilitated the transformation and upgrading of agricultural production methods, increasing farmers' income.

Currently, the area and yield of facility-grown strawberries are increasing annually, especially in North China, East China, and the Southwest region, where mature production bases have been established. With advancements in agricultural technology, cultivation techniques will further develop, leading to greater intelligence and precision and promoting the sustainable development of the strawberry industry.

6.1.2 Characteristics of Strawberry Cultivation in Southern Xinjiang

Southern Xinjiang, located in northwest China, has an arid to semi-arid climate with ample sunlight and significant diurnal temperature variations, which are favorable for strawberry growth. Adequate sunlight enhances photosynthesis, increasing fruit sweetness, while the large temperature range aids nutrient accumulation, improving quality. However, water resources are scarce, posing irrigation challenges. Traditional open-field irrigation is ineffective, so drip irrigation and integrated water-fertilizer technologies are used to precisely control water and nutrient supply, increasing efficiency and ensuring adequate moisture and nutrients. The soil in Southern Xinjiang is often sandy or gravelly, with poor water and nutrient retention. Soilless cultivation techniques, using substrates like coconut coir, perlite, or sand, address these issues, reducing soil-borne diseases and improving yield and quality. Facility agriculture in Southern Xinjiang is rapidly developing, with greenhouses effectively controlling temperature and humidity. This allows for early marketing of strawberries in winter or early spring, yielding higher economic returns.

Despite the climate advantages, challenges such as frost risk due to temperature fluctuations remain. Strawberry cultivation in Southern Xinjiang leverages the region's climate and resource advantages, successfully overcoming challenges and achieving strong competitiveness in quality and yield. With ongoing advancements, the strawberry industry in Southern Xinjiang is expected to bring greater economic benefits to local farmers.

6.1.3 Challenges in Water and Fertilizer Management for Strawberries

Water and fertilizer management are critical aspects of strawberry cultivation, significantly impacting fruit growth, yield, and quality. However, several challenges arise in the management process, especially in regions facing water scarcity and inefficient fertilizer use. These challenges not only complicate cultivation but also place increased demands on growers. In arid and semi-arid regions like Southern Xinjiang, water scarcity poses a primary challenge for strawberry cultivation. Strawberries require substantial water, particularly during the flowering, fruit-setting, and fruit-sizing stages. Insufficient water can directly diminish fruit quality and affect yield. Traditional irrigation methods often result in water wastage and may lead to excessive soil moisture around the roots, creating conditions conducive to pests and diseases. Thus, ensuring that strawberries receive adequate and appropriate amounts of water in such limited conditions is a pressing issue. Another significant challenge is low fertilizer use efficiency. Strawberries have high nutrient demands, particularly for nitrogen, phosphorus, and potassium. However, traditional fertilization methods often prove inefficient, leading to substantial fertilizer waste. The shallow root system of strawberries makes them especially prone to nutrient loss from over-fertilization, particularly during heavy rains or excessive irrigation. Nutrient runoff not only increases cultivation costs but also contributes to fertilizer waste and environmental problems, such as soil salinization and groundwater pollution. Uneven application of water and fertilizer also presents a major challenge. Strawberries have varying water and nutrient needs at different growth stages, and traditional management practices often struggle to meet these precise requirements. For instance, the fruit-sizing stage demands the highest levels of water and nutrients; inadequate supply during this critical period can adversely affect fruit size and quality. Conversely, if high-demand watering and fertilization practices continue into the later growth stages, it can result in fertilizer waste and excessive fruit growth, ultimately impacting market value.

Effective management requires addressing these issues through modern techniques and technologies to optimize water and fertilizer use, enhance crop yield and quality, and mitigate environmental impact.

Additionally, strawberries are highly sensitive to environmental conditions, with temperature, humidity, light, and soil conditions all significantly affecting their growth. Extreme weather events—such as droughts, high temperatures, and cold spells—heighten the water and nutrient demands of strawberries. However, these conditions often do not receive timely adjustments in traditional management practices. Over-irrigation or excessive fertilization can also lead to oxygen deficiency in the root system, heightening the risk of pests and diseases. Despite continuous advancements in agricultural technologies, many strawberry cultivation areas still rely on traditional methods and manual operations for water and fertilizer management, lacking scientific monitoring and control methods. Precision irrigation and fertilization technologies are not yet widely adopted in strawberry farming, and many farmers lack the necessary technical support. This often results in suboptimal water and fertilizer management. Even where drip irrigation or integrated water-fertilizer systems are available, their effectiveness can be limited without targeted and well-planned management strategies.

Water and fertilizer management for strawberries not only influences production but also affects the broader ecological environment. Excessive use of water and chemical fertilizers can lead to soil degradation and water quality pollution, disrupting the local ecological balance. In areas with limited water resources, the high water demands of strawberry cultivation can create conflicts with other local needs, intensifying competition for water. Moreover, poor water and fertilizer management directly increases the economic costs of strawberry cultivation. Water wastage and inefficient fertilizer use not only lead to immediate economic losses but also contribute to hidden costs associated with environmental management. In a competitive market, high production costs can directly impact growers' profitability. To address these challenges, effective water and fertilizer management for strawberries requires the adoption of modern agricultural technologies. Precision irrigation, drip irrigation systems, integrated water-fertilizer technologies, and smart agricultural systems are essential. By utilizing real-time monitoring of soil moisture, nutrient levels, and strawberry growth conditions, growers can accurately adjust water and fertilizer applications, ensuring optimal moisture and nutrient levels at each growth stage. Furthermore, promoting agricultural technology and providing farmer training to advance water and fertilizer management techniques are crucial for enhancing the sustainability and benefits of strawberry cultivation. Through scientific management and the application of cutting-edge technologies, issues related to water and fertilizer can be effectively managed. This approach leads to improved strawberry yield and quality, reduced production costs, and minimized environmental impact, ultimately fostering sustainable agricultural development.

6.2 Crop Evapotranspiration and Water-Fertilizer Demand Theory

6.2.1 Fundamental Theory of Evapotranspiration

Evapotranspiration is the combined process of water loss through both plant transpiration and the evaporation of moisture from soil surfaces. Transpiration involves plants releasing water vapor into the atmosphere through stomata on their leaves, while evaporation refers to the direct transformation of moisture from the soil or plant surfaces into water vapor. Evapotranspiration is vital in agriculture as it provides a crucial indicator of a crop's water needs and serves as essential reference data for irrigation management. By calculating and predicting evapotranspiration rates, farmers can precisely determine the water requirements of their crops at different times. This enables them to develop scientifically-based irrigation strategies that avoid both over-irrigation and under-irrigation, optimizing water use and supporting healthy crop growth.

Evapotranspiration reflects the actual water needs of crops and requires consideration of several factors, including temperature, humidity, wind speed, sunlight, plant type, and growth stage. These elements directly impact the rates of transpiration and evaporation, which in turn determine the level of evapotranspiration. When evapotranspiration rates are high, crops have increased water needs, necessitating additional irrigation to support their growth. Conversely, lower evapotranspiration rates indicate reduced water requirements, allowing for a reduction in irrigation to prevent waste. Properly adjusting irrigation based on evapotranspiration helps ensure efficient water use and supports optimal crop health and productivity.

In agricultural irrigation, evapotranspiration plays a crucial role in optimizing water use, enhancing crop yield and quality, mitigating environmental impact, and addressing climate change challenges. Accurate measurement and prediction of evapotranspiration enable farmers to tailor irrigation practices to the precise water needs of crops. This approach not only helps prevent water waste, especially in water-scarce regions, but also maximizes the efficiency of water resource utilization. Adequate moisture is essential for healthy crop growth and high yield. By aligning irrigation volumes with evapotranspiration rates, farmers ensure that crops receive the right amount of water during critical growth stages. This supports robust photosynthesis, efficient nutrient absorption, and ultimately boosts crop yield and quality. On the other hand, over-irrigation can waste water, contribute to soil salinization, cause nutrient loss, and lead to groundwater pollution. Managing irrigation based on evapotranspiration data helps mitigate these environmental issues, maintain soil health, and promote the sustainability of agricultural ecosystems. As climate change introduces uncertainties, evapotranspiration becomes a dynamic indicator of crop water needs, helping farmers adapt to climatic fluctuations. For instance, during droughts or high-temperature periods, increased evapotranspiration signals a need for more irrigation, allowing farmers to address water deficits and prevent crop stress. Therefore, evapotranspiration data is a vital reference in agricultural irrigation management.

It enables farmers to optimize irrigation strategies, improve efficiency, and support sustainable agricultural practices through informed decision-making.

6.2.2 *Application of the Penman–Monteith Formula*

The Penman–Monteith formula is a widely accepted tool in agriculture and water resource management for calculating reference evapotranspiration (ET_0), which represents the amount of evapotranspiration a crop needs under specific climatic conditions. This formula integrates principles of aerodynamics and energy balance, offering an accurate method for estimating evapotranspiration by incorporating various meteorological factors. Recommended by the Food and Agriculture Organization (FAO) as the standard method for determining crop water requirements, the Penman–Monteith formula provides a reliable basis for optimizing irrigation practices and managing water resources effectively.

The standard form of the Penman–Monteith formula (6.1) for calculating reference evapotranspiration (ET_0) is as follows:

$$ET_0 = \frac{0.408\Delta(R_n - G) + \gamma\frac{900}{T_{\text{mean}}+273}u_2(e_s - e_a)}{\Delta + \gamma(1 + 0.34u_2)} \tag{6.1}$$

The meanings and units of the various symbols are:

ET_0: Reference evapotranspiration, measured in millimeters per day (mm/day), indicating the daily amount of evapotranspiration.

Δ: Slope of the saturation vapor pressure curve, measured in kilopascals per degree Celsius (kPa/°C), representing the change in vapor pressure due to temperature changes.

R_n: Net radiation flux, measured in megajoules per square meter per day (MJ/m^2/day), indicating the net energy entering the crop canopy.

G: Soil heat flux density, measured in megajoules per square meter per day (MJ/m^2/day), indicating the change in soil heat. For daily scale calculations, it can be ignored and is typically taken as 0.

γ: Psychrometric constant, measured in kilopascals per degree Celsius (kPa/°C), a constant in the aerodynamic part of the formula dependent on atmospheric pressure.

T: Daily average air temperature, measured in degrees Celsius (°C), a key temperature parameter affecting evapotranspiration.

u_2: Wind speed at 2 m height, measured in meters per second (m/s), indicating the wind speed 2 m above the surface.

e_s: Saturation vapor pressure, measured in kilopascals (kPa), related to temperature and calculated based on the maximum moisture content in the air.

e_a: Actual vapor pressure, measured in kilopascals (kPa), calculated based on the actual moisture content in the air, reflecting the humidity of the air.

$e_s - e_a$: Saturation vapor pressure deficit, indicating the evaporation capacity of the air, depending on the differences in temperature and humidity.

Saturation vapor pressure can be determined from environmental parameters such as the daily maximum and minimum temperatures. It is the average of the saturation vapor pressures at the daily maximum and minimum temperatures. The calculation is given by the following formula (6.2):

$$e_s = \frac{0.6108 \times \exp\left[\frac{17.27T_{max}}{T_{max}+237.3}\right] + 0.6108 \times \exp\left[\frac{17.27T_{min}}{T_{min}+237.3}\right]}{2} \tag{6.2}$$

In formula (6.2), T_{max} represents the daily maximum temperature in degrees Celsius (°C), and T_{min} represents the daily minimum temperature in degrees Celsius (°C).

Actual vapor pressure can be derived from the average relative humidity, and its calculation method is given by formula (6.3).

$$e_a = \frac{100RH_{mean}}{e_s} \tag{6.3}$$

In formula (6.3), RH_{mean} represents the average relative humidity in percentage (%).

The calculation method for the slope of the saturation vapor pressure curve, Δ is given by formula (6.4).

$$\Delta = \frac{4098e_s}{(T_{mean} + 237.3)^2} \tag{6.4}$$

The calculation method for the psychrometric constant is given by formula (6.5).

$$\gamma = 0.665 \times 10^{-3} \times P \tag{6.5}$$

In formula (6.5), P represents the atmospheric pressure at the measurement point, in kilopascals (kPa).

The reference evapotranspiration (ET_0) calculated using the Penman–Monteith formula is crucial for effective agricultural irrigation. ET_0 represents the amount of water a crop needs under optimal conditions, allowing farmers and irrigation managers to precisely determine daily irrigation requirements. This method is particularly valuable for crops that require meticulous water management, such as strawberries and grapes in controlled environments. By integrating regional meteorological data with crop physiological characteristics, the Penman–Monteith formula provides a solid scientific foundation for irrigation practices. It aids in optimizing water resource use, improving crop yield and quality, and minimizing resource wastage and environmental impacts associated with over- or under-irrigation.

Overall, the Penman–Monteith formula is an indispensable tool in modern agricultural irrigation management. By delivering accurate evapotranspiration calculations, it empowers farmers and water resource managers to fine-tune their irrigation strategies, ensuring optimal crop growth and fostering sustainable production practices.

6.2.3 *Construction and Optimization of Evapotranspiration Models*

6.2.3.1 Overview of Evapotranspiration Model Experiments

The experiment was conducted in Greenhouse No. 3 at the Yingxia Township Facility Agriculture Industrial Park in Korla City (As shown in Figs. 6.1 and 6.2), located at 41°46′13″ N latitude and 86°5′18″ E longitude. This greenhouse, measuring 75 m in length and 12 m in width, is oriented east–west and features a full steel frame constructed with advanced materials. It includes both upper and lower ventilation openings and is equipped with 40-mesh screens to protect against insects. In this facility, strawberries were cultivated using a container-based system, employing canvas substrate bags that each measure 20 cm in diameter and height. Each bag contained two strawberry seedlings, with a total of 100 rows used in the experiment. The "Christmas Red" variety, propagated locally through runners, was chosen for its suitability to the local climate, high-quality fruit production, strong disease resistance, and significant economic value. The facility collected relevant environmental data and actual evapotranspiration measurements. Through experimentation and data analysis, the optimized evapotranspiration model significantly enhanced calculation accuracy, providing a scientific foundation for precise irrigation of facility-grown strawberries.

Fig. 6.1 Experimental site 1

Fig. 6.2 Experimental site 2

6.2.3.2 Experimental Methods

From November 1, 2023, to December 30, 2023, environmental parameters within the greenhouse were continuously monitored for a period of 60 days. Daily maximum and minimum temperatures, average temperature, maximum and minimum humidity, average humidity, and sunshine duration were recorded using a small weather station, the IWS3, from Guangzhou Xfly Technology Co., Ltd. Daily atmospheric pressure data were sourced from the China Meteorological Administration. Solar net radiation was measured with a TBB-1L solar net radiation meter. Additionally, the elevation and geographic coordinates of the test site were obtained through online mapping services such as Baidu Maps.

Once the strawberries had completed their acclimatization period, measurements of actual evapotranspiration began. Nine well-developed strawberry pots were uniformly selected within the greenhouse. Each day at 10:00 AM, the weight of each pot was recorded using the JSB15-01 electronic scale from Shanghai Puchun Measuring Instruments Co., Ltd., which has a capacity of 15 kg and an accuracy of 0.5 g. The actual evapotranspiration was determined by calculating the weight difference between two consecutive measurements. The average evapotranspiration value from the nine pots was then used as the final measurement for actual evapotranspiration.

The Penman–Monteith formula was implemented in MATLAB 2022b to calculate theoretical evapotranspiration by substituting the collected environmental parameters into Eq. (6.1). Initial data processing and recording were carried out using Excel 2016. A fit analysis of the theoretical and actual evapotranspiration was performed using Minitab 19 software, and graphs were generated using Origin 2018b software.

Experimental data reveal that strawberry evapotranspiration varies across different growth stages and is closely linked to environmental conditions. A comparison of the actual measured evapotranspiration with theoretical values calculated using the Penman–Monteith formula showed some discrepancies, though the overall trends

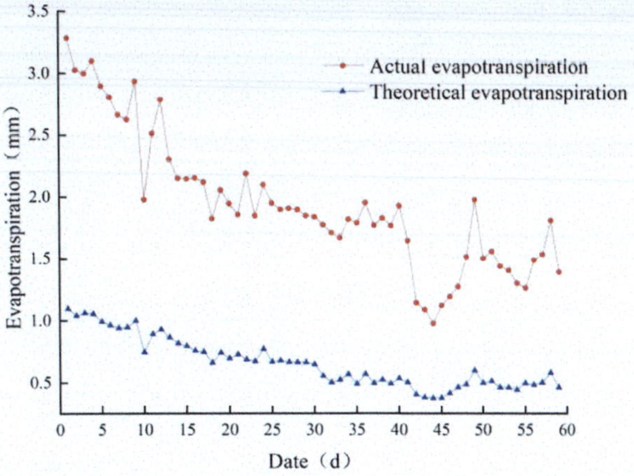

Fig. 6.3 Actual versus theorctical evapotranspiration of facility-grown strawberries

were consistent. To enhance the model's predictive accuracy, we optimized the theoretical model by adjusting the weights of parameters such as temperature, humidity, and radiation. This refinement aligned the optimized model more closely with the actual measurement results.

The results for actual and theoretical evapotranspiration from November 1, 2023, to December 30, 2023, are illustrated in Fig. 6.3.

On November 1, 2023, the actual evapotranspiration (ET) for facility-grown strawberries peaked at 3.283 mm. Conversely, on December 15, it dropped to a minimum of 0.981 mm. The decline in ET around November 15 was attributed to snowfall in Korla City, which prevented the greenhouse blankets from being raised and consequently reduced evapotranspiration. During this period, the theoretical ET, calculated using the Penman–Monteith formula, ranged from a maximum of 1.098 mm on November 1 to a minimum of 0.378 mm on December 15. The observed discrepancy between actual and theoretical ET values can be attributed to the limitations of the Penman–Monteith model, which is based on a single-layer, large-leaf assumption that is more suited for dense vegetation. In this experiment, strawberry leaves did not fully cover the ground, and the presence of black weed cloth further influenced the lower theoretical ET values. Despite this, the overall trends in actual and theoretical ET were consistent, indicating a good correlation and validity of the model. To enhance accuracy, the Penman–Monteith formula was optimized by adjusting the parameters to better fit the actual evapotranspiration data. The fitting results of the theoretical ET to the actual ET are presented in Fig. 6.4, while Fig. 6.5 illustrates the comparison of model errors before and after optimization. The optimized evapotranspiration model for facility-grown strawberries, based on the Penman–Monteith formula, is given by Eq. (6.6).

$$ET_c = 0.2049 + 2.632ET_0(R^2 = 0.93) \tag{6.6}$$

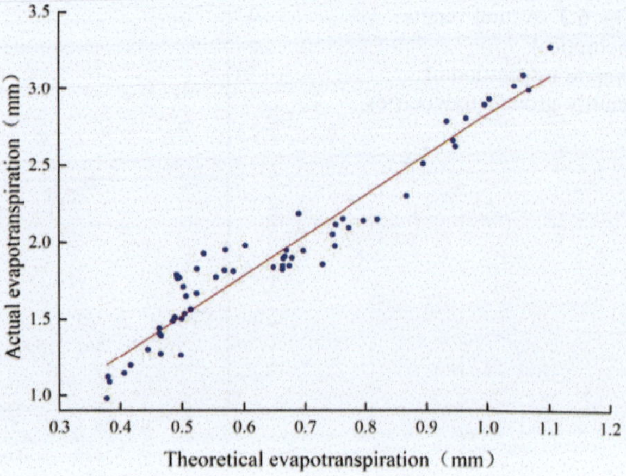

Fig. 6.4 Fitting of actual evapotranspiration and theoretical evapotranspiration

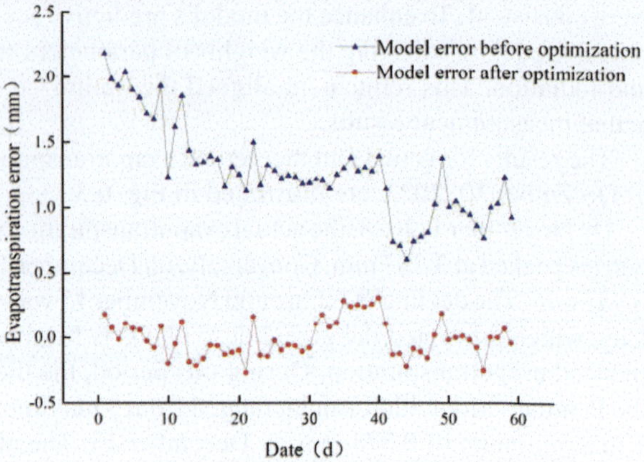

Fig. 6.5 Comparison of errors before and after model optimization

As depicted in Fig. 6.5, the model errors before optimization were notably large, with a maximum error of 2.185 mm and a minimum error of 0.603 mm, resulting in an average absolute error (MAD) of 1.279 mm. Following optimization, the errors were significantly reduced, with the maximum error dropping to 0.313 mm and the minimum error to 0.007 mm. The average absolute error (MAD) was reduced to 0.117 mm, with a standard deviation (S) of 0.145 and a mean square error (MSE) of 2.2045. Compared to the pre-optimization model, the accuracy of the optimized model improved substantially, demonstrating an average error of just 6.05% relative to the measured values.

6.3 Water and Fertilizer Management Model for Facility Strawberries

6.3.1 Basic Concepts of Water and Fertilizer Management

Water and fertilizer management involves the precise allocation and supply of water and nutrients tailored to the specific needs of crops throughout their growth and development. This approach integrates knowledge from crop physiology, hydrology, soil science, and plant nutrition, with the goal of delivering the appropriate amounts of water and nutrients at various growth stages. By optimizing these inputs, this technique aims to enhance crop growth, increase yields, and improve overall quality.

During crop growth, water and nutrient requirements vary significantly at different stages. Generally, crops need less water and fewer nutrients in the early stages of growth, but these needs increase substantially during flowering and fruit enlargement. Consequently, precise water and fertilizer management is essential to match the specific needs of each growth stage, ensuring that crops receive the optimal amounts of water and nutrients at the right times. The soil–plant-atmosphere continuum theory explains the transfer of water and nutrients between the soil, plants, and atmosphere. Plants absorb water and nutrients from the soil through their roots and release water into the atmosphere via transpiration. Effective water and fertilizer management must ensure that the soil maintains adequate moisture and nutrient levels to meet the plant's absorption needs. Water and nutrients are interrelated; suitable moisture conditions enhance nutrient uptake, while adequate nutrient supply can improve a crop's drought resistance. Thus, water and fertilizer management involves the coordinated management of both resources to maximize their synergistic benefits for crop growth. In facility agriculture, advanced water and fertilizer management technologies, such as drip irrigation and micro-spray systems, are commonly employed. These systems deliver water and nutrients directly to the crop roots, minimizing waste and preventing nutrient loss from the soil, thereby enhancing the efficiency of water and fertilizer use. Such technologies are particularly effective for high-value crops like facility-grown strawberries, tomatoes, and cucumbers. Water-fertilizer integration technology involves dissolving water and fertilizer together and delivering them simultaneously to the crop roots through the irrigation system. This technology allows for automatic adjustments of water and fertilizer supplies based on the crop's real-time needs, ensuring the optimal water-to-fertilizer ratio at various growth stages. As a result, it improves crop growth rates, yield, and quality.

6.3.2 Factors Influencing Water and Fertilizer Needs

Strawberries, being a high-value horticultural crop, have specific water and fertilizer needs that are influenced by several factors, including the type of fertilizer used, the amount of irrigation, and prevailing environmental conditions. A comprehensive understanding of these factors is essential for optimizing the growth environment, enhancing yield, and improving fruit quality.

Firstly, the type of fertilizer used significantly impacts the water and nutrient needs of strawberries. Nitrogen is crucial for promoting leaf and stem growth, which increases transpiration and water demand. However, excessive nitrogen application can lead to excessive vegetative growth, decreased fruit quality, and soil salinization, ultimately affecting water and fertilizer requirements. Phosphorus is essential for root development and fruit formation; adequate phosphorus enhances the root system's ability to absorb water and nutrients, potentially reducing overall water needs. Conversely, excessive phosphorus can accumulate in the soil, disrupting nutrient balance. Potassium improves drought resistance and disease resilience in strawberries. Proper potassium application enhances water use efficiency and reduces evaporation. Insufficient potassium, on the other hand, can decrease drought resistance and increase water requirements. Organic fertilizers contribute to improved soil structure, increased water retention, and enhanced fertility, which can reduce irrigation needs. However, in drought conditions, the decomposition of organic fertilizers may require additional water, which must be accounted for in water management strategies.

Irrigation amount is another crucial factor influencing the water and fertilizer needs of strawberries. Over-irrigation can saturate the soil, impairing root respiration and nutrient absorption, and may lead to nutrient leaching, thereby increasing the need for additional fertilizer. Proper irrigation, however, maintains optimal soil moisture levels, ensuring effective water and nutrient uptake by the roots. This not only maximizes the efficiency of water and fertilizer use but also minimizes nutrient loss and reduces the risk of soil salinization. Conversely, insufficient irrigation can cause soil dryness, hindering water and nutrient absorption, stunting growth, and diminishing yield and quality. This condition often results in an increased need for both water and fertilizer to compensate for the deficiencies.

Environmental conditions, such as temperature, humidity, light intensity, and wind speed, also significantly affect strawberry water and fertilizer needs. High temperatures increase transpiration and water demand, and enhance nutrient absorption and utilization. In contrast, low temperatures slow down growth and reduce water and fertilizer needs. Air humidity directly affects the rate of transpiration. Higher humidity can reduce transpiration, lowering water needs but possibly increasing disease risks; lower humidity accelerates transpiration, increasing water needs and requiring more frequent irrigation. Light intensity affects photosynthesis, thereby influencing water and nutrient needs. Adequate light supports growth, improving water and fertilizer use efficiency, whereas insufficient light reduces photosynthesis efficiency and decreases water and fertilizer needs. Wind speed is also a significant factor; higher wind speeds increase water evaporation from leaf surfaces, accelerating transpiration and increasing water needs. Additionally, high wind speeds can exacerbate soil moisture evaporation, further increasing irrigation requirements.

The water and fertilizer needs of strawberries are influenced by a combination of factors, including environmental conditions and crop requirements. Effective strawberry cultivation demands scientifically informed water and fertilizer management strategies tailored to these specific conditions. To achieve optimal results, it is crucial to develop and dynamically adjust management plans based on real-time environmental data and crop needs. Ensuring that strawberries receive the appropriate amounts of water and nutrients at each growth stage allows for resource optimization. This approach not only enhances yield and fruit quality but also supports efficient and sustainable production practices. By integrating precise water and fertilizer management, growers can meet both agricultural goals and environmental stewardship.

6.3.3 *Establishment of the Water and Fertilizer Management Model*

The experimental area details remain as previously described. During the flowering period, bees from the Bazhou Bee Breeding Farm were used to pollinate the strawberries. The bee hives were placed towards the back wall in the middle of the greenhouse.

The irrigation water used in the experiment was sourced from local groundwater, characterized by an electrical conductivity of 1.43 mS/cm and a pH of 7.1. The fertilizers applied included urea from Hubei Yihua Group Co., Ltd. (46% N), superphosphate from Tianjin Zhiyuan Chemical Reagent Co., Ltd. (16% P_2O_5), potassium sulfate from Guotou Xinjiang Lop Nur Potash Co., Ltd. (52% K_2O), and monoammonium phosphate from Yunnan Zhongzheng Chemical Industry Co., Ltd. (12% N, 61% P_2O_5). All fertilizers were standardized to their effective nutrient content for the experiment.

The planting substrate for the experiment was yellow sand sourced from the Taklamakan Desert. Prior to planting, the substrate was leached with irrigation water to reduce its salt and alkali content. Additionally, the substrate was sterilized using a 2‰ concentration of fludioxonil and a 1‰ concentration of propiconazole. The physical and chemical properties of the yellow sand substrate were assessed using the ring knife method and the specific gravity bottle method, as detailed in Table 6.1.

The experimental variety and planting containers were as previously described. Strawberry plants were pruned to retain only two leaves and one heart per plant. Planting commenced on September 31, 2023, following the guidelines of "deep but not burying the roots, shallow but not exposing the roots, and arching the back outward."

Table 6.1 Physical and chemical properties of yellow sand substrate

Bulk density	Electrical conductivity	pH value	Field capacity	Soil porosity	Air-filled porosity	Water-holding porosity
1.489	0.35 mS/cm	6.5–7.2	22.9%	43%	8.9%	34.1%

Fig. 6.6 Planting condition of facility strawberries

Following planting, the following treatments were applied: Disease Prevention: A 1% concentration of metalaxyl-mancozeb fungicide from Shanghai Hebon Pharmaceutical Co., Ltd. was sprayed on the leaves to prevent downy mildew. Promoting Growth: A 0.5% concentration of gibberellic acid and thiourea from Weifang Wansheng Biological Pesticide Co., Ltd. was applied to break strawberry dormancy and encourage root development. Soil Disease Control: A 1% concentration of fludioxonil from Syngenta Nantong Crop Protection Co., Ltd. was used for root drenching to prevent soil-borne diseases. During the acclimatization period, cotton blankets were used for shading from 13:30 to 16:30 daily. Ventilation openings were left open at night to reduce temperatures and support flower bud differentiation. The water and fertilizer coupling experiment began on November 15, 2023, and concluded in early March 2024. The planting situation of the facility strawberries is illustrated in Fig. 6.6.

To investigate the impact of different fertilization amounts on strawberry yield under uniform planting conditions, the experiment evaluated four key factors: irrigation amount, nitrogen fertilizer application, phosphorus fertilizer application, and potassium fertilizer application. Each factor was tested at five levels, resulting in a four-factor, five-level quadratic regression composite design (RCCD) with a total of 30 treatment groups. To maintain a consistent greenhouse environment throughout the experiment, the central six adjacent rows in the greenhouse were selected for testing. Each row was divided into five treatment groups arranged alternately, with two repetitions for each treatment. This setup ensured robust and reliable results. The schematic diagram of the experimental planting area is illustrated in Figs. 6.7 and 6.8.

Based on the optimized Penman–Monteith evapotranspiration model, the optimal daily water requirement for the crop was established. This optimal water requirement served as the baseline (zero level) for irrigation in the quadratic regression composite design experiment. During the experiment, irrigation frequency was adjusted in response to fluctuations in evapotranspiration. Consequently, the irrigation volume varied between an upper limit of 57.118 m^3 per 667 m^2 and a lower limit of 37.154 m^3 per 667 m^2.

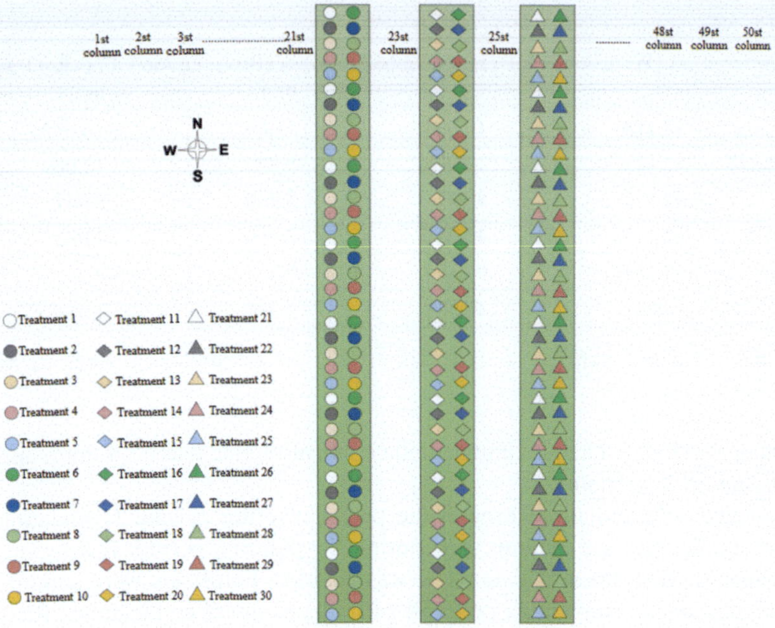

Fig. 6.7 Schematic diagram of the experimental planting area

Fig. 6.8 Actual view of the experimental area

Based on the preliminary fertilization experiment results, the following fertilizer application rates were established:

Nitrogen (N): The maximum single application rate during the leaf expansion period was 41.4 mg/plant, while the minimum was 0 mg/plant.

Phosphorus (P_2O_5): The maximum single application rate was 30.52 mg/plant, with the minimum also being 0 mg/plant.

Table 6.2 Natural factors level coding table (r = 2)

$z_j(x_j)$	z_1 (W: $m^3/667\ m^2$)	z_2 (N: $kg/667\ m^2$)	z_3 (P_2O_5: kg /667 m^2)	z_4 (K_2O: kg/667 m^2)
$z_{2j}(2)$	57.118	16.093	11.862	16.093
$z_{0j} + \Delta_j(1)$	52.127	12.071	8.896	12.069
$z_{0j}(0)$	47.136	8.047	5.930	8.047
$z_{0j} - \Delta_j(-1)$	42.145	4.024	2.966	4.022
$z_{1j}(-2)$	37.154	0	0	0
$\Delta_j = \frac{z_{2j}-z_{1j}}{4}$	4.991	4.023	2.966	4.023
$x_j = \frac{z_j-z_{0j}}{\Delta_j}$	$x_1 = \frac{z_1-47.136}{4.991}$	$x_2 = \frac{z_2-8.047}{4.023}$	$x_3 = \frac{z_3-5.930}{2.996}$	$x_4 = \frac{z_4-8.047}{4.023}$

Potassium (K_2O): The maximum single application rate was 41.4 mg/plant, with a minimum of 0 mg/plant.

From bud differentiation through the peak flowering period, fertilization was applied at 1.5 times the amount used during the planting period. During the fruit enlargement period, the application rate was doubled. Fertilizers were administered via irrigation water following a principle of frequent, small applications. Real-time monitoring of nutrient content guided the adjustment of fertilization frequency. The total fertilizer application rates per unit area for the entire experimental cycle were determined as follows:

Nitrogen (N): Upper limit: 16.093 kg/667 m^2; Lower limit: 0 kg/667 m^2.

Phosphorus (P_2O_5): Upper limit: 11.862 kg/667 m^2; Lower limit: 0 kg/667 m^2.

Potassium (K_2O): Upper limit: 16.093 kg/667 m^2; Lower limit: 0 kg/667 m^2.

The natural factors level codes are presented in Table 6.2, and the experimental structure matrix is detailed in Table 6.3.

6.3.4 *Experimental Analysis: Results of the Water and Fertilizer Management Model*

After the strawberries have ripened, the fruits are collected every three days. Once all the flowers have fruited, the yield for each of the 30 treatments is recorded and then converted to yield per 667 m^2 based on the treatment area. The results for each treatment are detailed in Table 6.4.

6.3.4.1 Establishment of the Regression Model

Using yield as the final evaluation metric, a regression model for the interactions among irrigation volume, nitrogen, phosphorus, and potassium was established using DPS software, as shown in Eq. (6.7).

Table 6.3 Structure matrix for quadratic general rotational combination test design

处理	因素			
	X_1 (W: $m^3/667\ m^2$)	X_2 (N: kg/667 m^2)	X_3 (P_2O_5: kg/667 m^2)	X_4 (K_2O: kg/667 m^2)
T1	1 (52.127)	1 (12.071)	1 (8.896)	1 (12.069)
T2	1 (52.127)	1 (12.071)	1 (8.896)	−1 (4.022)
T3	1 (52.127)	1 (12.071)	−1 (2.966)	1 (12.069)
T4	1 (52.127)	1 (12.071)	−1 (2.966)	−1 (4.022)
T5	1 (52.127)	−1 (4.024)	1 (8.896)	1 (12.069)
T6	1 (52.127)	−1 (4.024)	1 (8.896)	−1 (4.022)
T7	1 (52.127)	−1 (4.024)	−1 (2.966)	1 (12.069)
T8	1 (52.127)	−1 (4.024)	−1 (2.966)	−1 (4.022)
T9	−1 (42.145)	1 (12.071)	1 (8.896)	1 (12.069)
T10	−1 (42.145)	1 (12.071)	1 (8.896)	−1 (4.022)
T11	−1 (42.145)	1 (12.071)	−1 (2.966)	1 (12.069)
T12	−1 (42.145)	1 (12.071)	−1 (2.966)	−1 (4.022)
T13	−1 (42.145)	−1 (4.024)	1 (8.896)	1 (12.069)
T14	−1 (42.145)	−1 (4.024)	1 (8.896)	−1 (4.022)
T15	−1 (42.145)	−1 (4.024)	−1 (2.966)	1 (12.069)
T16	−1 (42.145)	−1 (4.024)	−1 (2.966)	−1 (4.022)
T17	−2 (37.154)	0 (8.047)	0 (5.930)	0 (8.047)
T18	2 (57.118)	0 (8.047)	0 (5.930)	0 (8.047)
T19	0 (47.136)	−2 (0)	0 (5.930)	0 (8.047)
T20	0 (47.136)	2 (16.093)	0 (5.930)	0 (8.047)
T21	0 (47.136)	0 (8.047)	−2 (0)	0 (8.047)
T22	0 (47.136)	0 (8.047)	2 (11.862)	0 (8.047)
T23	0 (47.136)	0 (8.047)	0 (5.930)	−2 (0)
T24	0 (47.136)	0 (8.047)	0 (5.930)	2 (16.093)
T25	0 (47.136)	0 (8.047)	0 (5.930)	0 (8.047)
T26	0 (47.136)	0 (8.047)	0 (5.930)	0 (8.047)
T27	0 (47.136)	0 (8.047)	0 (5.930)	0 (8.047)
T28	0 (47.136)	0 (8.047)	0 (5.930)	0 (8.047)
T29	0 (47.136)	0 (8.047)	0 (5.930)	0 (8.047)
T30	0 (47.136)	0 (8.047)	0 (5.930)	0 (8.047)

$$\begin{aligned} Y = {} & 1306.99 + 24.45x_1 + 38.56x_2 + 25.81x_3 + 22.55x_4 - 13.01x_1^2 - 16.84x_2^2 \\ & - 32.38x_3^2 - 25.02x_4^2 + 19.17x_1x_2 - 10.00x_1x_3 + 22.87x_1x_4 + 14.54x_2x_3 \\ & - 19.59x_2x_4 + 17.66x_3x_4 \end{aligned} \tag{6.7}$$

Table 6.4 Experimental results

Treatment	Yield	Treatment	Yield	Treatment	Yield	Treatment	Yield
T1	1379.26	T9	1289.62	T17	1231.36	T25	1259.19
T2	1273.41	T10	1252.63	T18	1286.32	T26	1324.24
T3	1334.88	T11	1055.27	T19	1170.49	T27	1336.27
T4	1237.92	T12	1236.08	T20	1316.59	T28	1329.76
T5	1256.53	T13	1231.52	T21	1133.73	T29	1257.02
T6	1148.46	T14	1125.43	T22	1228.98	T30	1336.66
T7	1219.12	T15	1191.91	T23	1183.35	–	–
T8	1131.33	T16	1121.47	T24	1238.22	–	–

6.3.4.2 Regression Model Validation

From the F-distribution table, we have F0.05 (14, 16) = 2.37 and F0.05 (10, 6) = 4.06. The DPS software computed F = 6.902 > F0.05 (14, 16), and Flf = 1.526 < F0.05 (10, 6). With a regression equation correlation coefficient R = 0.92, this indicates that the model is well-fitting and statistically significant at $\alpha = 0.05$, with a confidence level of 95%.

To evaluate the impact of different coefficients on the metrics, t-tests were conducted on the regression model coefficients. For f = 6, the critical t-values are t0.1 = 1.943, t0.05 = 2.447, t0.01 = 3.707, denoted by *, **, and ***, respectively. The results indicate that irrigation volume x_1, nitrogen application x_2, phosphorus application x_3, and potassium application x_4 all have a significant impact on yield. The detailed results of these calculations are presented in Table 6.5.

After removing the insignificant terms, the regression model within the coded space was simplified, as shown in Eq. (6.8):

$$\begin{aligned} Y = {} & 1306.99 + 24.45x_1 + 38.56x_2 + 25.81x_3 + 22.55x_4 - 13.01x_1^2 - 16.84x_2^2 \\ & - 32.38x_3^2 - 25.02x_4^2 + 19.17x_1x_2 + 22.87x_1x_4 - 19.59x_2x_4 + 17.66x_3x_4 \end{aligned} \tag{6.8}$$

6.3.4.3 Main Factor Analysis

Since the coding and natural variables involve dimensionless linear transformations, the partial regression model remains unaffected by the absolute values and magnitudes of the factors. Thus, the partial regression coefficients directly reflect the impact of each factor on the yield. Based on the t-test results for the regression coefficients *b*, the order of impact of the experimental factors on the yield is as follows: nitrogen application amount (x_2) > phosphorus application amount (x_3) > irrigation amount

Table 6.5 Significant test of regression model coefficients b

Coefficient b	Test value t	Coefficient b	Test value t	Coefficient b	Test value t	Coefficient b	Test value t	Coefficient b	Test value t
b1	5.767***	b4	5.319***	b7	4.950***	b10	1.144	b13	2.241*
b2	9.095***	b5	1.989*	b8	3.825***	b11	2.616**	b14	2.020*
b3	6.088***	b6	2.574**	b9	2.193*	b12	1.663	–	–

Table 6.6 Factor levels for maximum yield

Factor	Irrigation amount (m^3/667 m^2)	Nitrogen application amount (kg/667 m^2)	Phosphorus application amount (kg/667 m^2)	Potassium application amount (kg/667 m^2)	Yield (kg/667 m^2)
Level	2	2	0	1	1446.47
Quota	57.118	16.093	5.930	12.069	

(x_1) > potassium application amount (x_4). The factor levels that maximize the yield are summarized in Table 6.6.

6.3.4.4 Single-Factor Analysis

By fixing three of the four influencing factors in the regression model at zero level, the partial regression sub-models for each individual factor affecting yield can be obtained. The models are given by Eqs. (6.9), (6.10), (6.11), and (6.12).

The partial regression sub-model for irrigation amount on yield is:

$$Y_w = 1306.99 + 24.45x_1 - 13.01x_1^2 \tag{6.9}$$

Partial Regression Sub-model for Nitrogen Fertilizer Application on Yield:

$$Y_N = 1306.99 + 38.56x_2 - 16.84x_2^2 \tag{6.10}$$

Partial Regression Sub-model for Phosphorus Fertilizer Application on Yield:

$$Y_w = 1306.99 + 25.81x_3 - 32.38x_3^2 \tag{6.11}$$

Partial Regression Sub-model for Potassium Fertilizer Application on Yield:

$$Y_K = 1306.99 + 22.55x_4 - 25.02x_4^2 \tag{6.12}$$

In the equations: Y_w, Y_N, Y_w, and Y_K represent the estimated yield values (in kg/667 m^2) when considering irrigation amount (x_1), nitrogen fertilizer application (x_2), phosphorus fertilizer application (x_3), and potassium fertilizer application (x_4) individually, respectively. The corresponding factor coding has upper and lower limits of −2 and 2. Based on the partial regression sub-models for each factor's effect on yield, the corresponding curve effect plots can be obtained, as shown in Fig. 6.9.

Based on Eqs. (6.9), (6.10), (6.11), and (6.12), the maximum values of the partial regression sub-models Y_w, Y_N, Y_P, and Y_K are 1312.1, 1311.45, 1308.27, and 1308.55, respectively, corresponding to factor values $x_1 = 0.938$, $x_2 = 1.144$, $x_1 = 0.399$, and $x_4 = 0.450$.

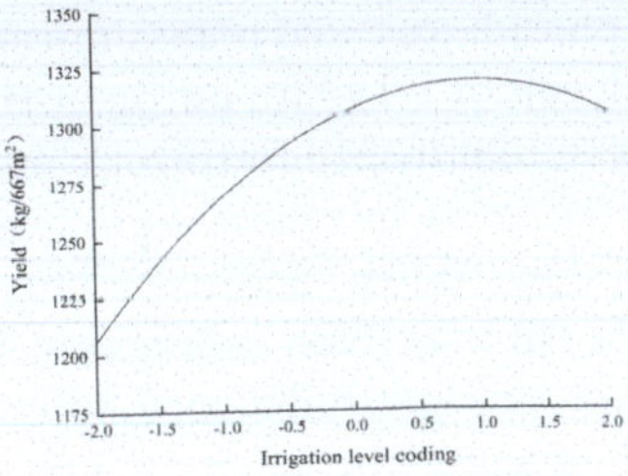

a: Relationship between Irrigation Amount and Yield

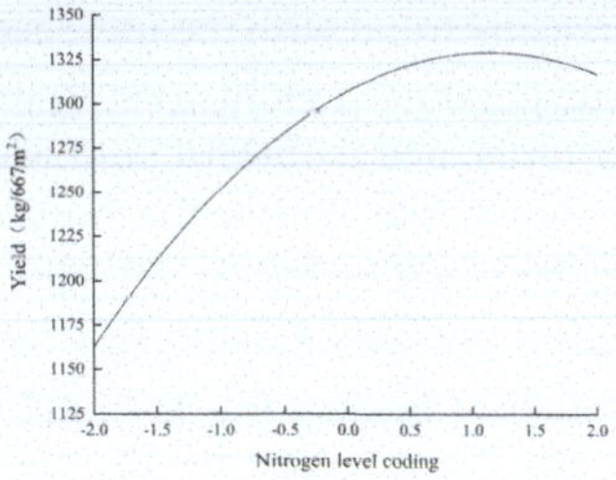

b: Relationship between Nitrogen Fertilizer and Yield

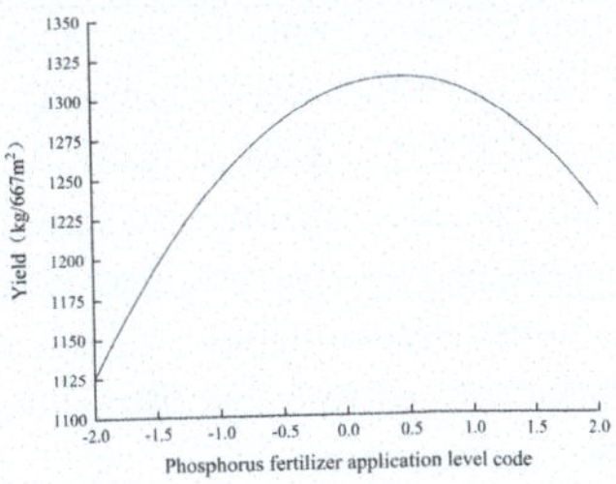

c:Relationship between Phosphorus Fertilizer and Yield

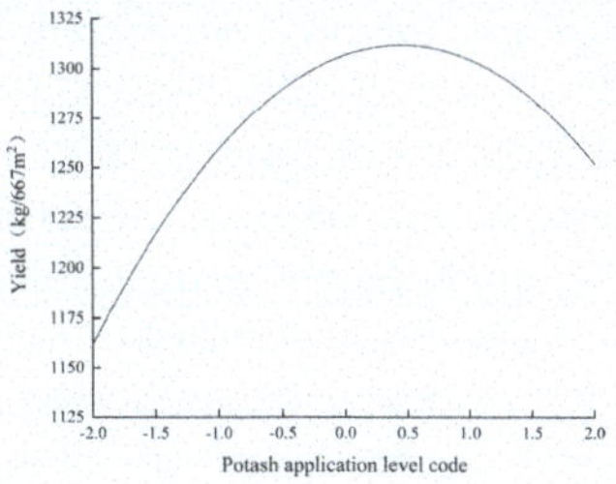

d:Relationship between Potassium Fertilizer and Yield

Fig. 6.9 Relationship between various factors and strawberry yield

From Fig. 6.9a–d, it is evident that the relationship curves between the four factors and yield, within the level range of −2 to 2, exhibit a parabolic shape that opens downwards. Figure 6.9a illustrates that the yield increases with irrigation up to a certain point, then begins to decrease. Specifically, in the irrigation range from −2 to −0.5, yield increases rapidly. However, as irrigation levels continue to rise, the rate of increase in yield slows down. The maximum yield is achieved when the irrigation level is at 0.938, with an estimated yield of 1312.1 kg/667 m^2. This trend can be attributed to the role of water as a carrier for nutrient absorption by plants. When water is insufficient, nutrient absorption efficiency decreases significantly, and water stress can inhibit growth, slow flower and fruit development, reduce transpiration, and negatively affect photosynthesis, ultimately decreasing yield. Conversely, excessive irrigation can lead to oxygen deficiency in the root zone due to the high bulk density and low aeration porosity of the yellow sand substrate. This deficiency can inhibit root growth and nutrient absorption. Additionally, the weak nutrient absorption capacity of the shallow strawberry roots and nutrient loss from the substrate can further reduce yield. Figure 6.9b–d show similar parabolic trends for nitrogen, phosphorus, and potassium applications, respectively, with yields initially increasing and then decreasing as the levels of these factors increase beyond an optimal point. These observations underscore the importance of balancing each factor to optimize strawberry yield. From Fig. 6.9b–d, it can be seen that the yield also follows an increasing then decreasing trend with the increase of nitrogen, phosphorus, and potassium application. The maximum yield is estimated at nutrient levels of 1.144 for nitrogen, 0.399 for phosphorus, and 0.450 for potassium. The analysis of nitrogen fertilizer's impact on yield shows that within the range of −2 to 0, the yield increases significantly. However, between 0 and 1.144, the yield growth slows down, and exceeding the optimal level not only fails to further increase yield but may even suppress it. This indicates that excessive nitrogen application leads to nutrient imbalance, affecting plant growth. For phosphorus, the yield increases significantly between −2 and −0.5 but slows down thereafter. Beyond the maximum application rate, additional phosphorus application results in a noticeable decrease in yield, indicating that excessive phosphorus inhibits root growth and affects plant development. For potassium, the yield increases notably between −2 and −0.4, but growth slows from −0.4 to 0.45, and further application results in a decrease in yield. Excessive potassium can disrupt root osmotic pressure, affecting water and nutrient absorption, and reducing yield. The single-factor analysis of nitrogen, phosphorus, and potassium reveals that strawberries have threshold levels for these nutrients. Within these thresholds, increasing fertilizer application enhances yield, but the rate of increase diminishes. Exceeding these thresholds can inhibit plant growth and decrease yield. Thus, irrigation, nitrogen, phosphorus, and potassium all contribute to plant growth. Within reasonable thresholds, appropriate increases in irrigation and fertilizer application can effectively enhance strawberry yield.

6.3.5 Determination of the Optimal Water and Fertilizer Allocation Scheme

Through the significance test of the regression model coefficients *b*, it was observed that all four factors affecting strawberry yield exhibit significant interactions. To analyze these interactions, we investigated the effects between pairs of factors while setting the other influencing factors to zero. As a result, binary quadratic sub-models were derived to capture the interactions between two factors at a time. These models provide insights into how the combined influence of two factors affects strawberry yield, further elucidating the complexity of the interactions in the experimental setup.

The interaction equation between irrigation amount (x_1) and nitrogen application (x_2) is given by Eq. (6.13).

$$Y = 1306.99 + 24.45x_1 + 38.56x_2 - 13.01x_1^2 - 16.84x_2^2 + 19.17x_1x_2 \tag{6.13}$$

The interaction equation between irrigation amount (x_1) and phosphorus application (x_3) is given by Eq. (6.14).

$$Y = 1306.99 + 24.45x_1 + 25.81x_3 - 13.01x_1^2 - 32.38x_3^2 \tag{6.14}$$

The interaction equation between irrigation amount (x_1) and potassium application (x_4) is given by Eq. (6.15).

$$Y = 1306.99 + 24.45x_1 + 22.55x_4 - 13.01x_1^2 - 25.02x_4^2 + 22.87x_1x_4 \quad (6.15)$$

The interaction equation between nitrogen application (x_2) and phosphorus application (x_3) is given by Eq. (6.16).

$$Y = 1306.99 + 38.56x_2 + 25.81x_3 - 16.84x_2^2 - 32.38x_3^2 \quad (6.16)$$

The interaction equation between nitrogen application (x_2) and potassium application (x_4) is given by Eq. (6.17).

$$Y = 1306.99 + 38.56x_2 + 22.55x_4 - 16.84x_2^2 - 25.02x_4^2 - 19.59x_2x_4 \quad (6.17)$$

The interaction equation between phosphorus application (x_3) and potassium application (x_4) is given by Eq. (6.18).

$$Y = 1306.99 + 25.81x_3 + 22.55x_4 - 32.38x_3^2 - 25.02x_4^2 + 17.66x_3x_4 \quad (6.18)$$

Based on the calculations, the interactions between the two factors are presented in Table 6.7a–f.

(a) From Table 6.7a and Fig. 6.10a, it is observed that with a nitrogen fertilizer application level of −2 (indicating no nitrogen applied), strawberry yield decreases as irrigation amount increases. However, when nitrogen fertilizer is applied, the yield increases with the irrigation amount. As the nitrogen level rises, the rate of yield increase also improves, peaking when both irrigation and nitrogen levels are at 2.
(b) From Table 6.7b and Fig. 6.10b, it can be seen that with a constant phosphorus fertilizer level, strawberry yield first increases and then decreases with rising irrigation amounts, reaching its maximum when irrigation is zero. Conversely, at a fixed irrigation amount, increasing phosphorus fertilizer initially boosts yield but later reduces it, with the highest yield occurring when both irrigation and phosphorus levels are at 0.
(c) From Table 6.7c and Fig. 6.10c, the relationship between potassium fertilizer application and strawberry yield shows that yield increases initially and then decreases as potassium levels rise, peaking at a potassium level of 1. Without potassium fertilizer, yield decreases with increasing irrigation. When potassium fertilizer is applied, yield increases with irrigation, and the effect is more pronounced with higher potassium levels. Maximum yield is achieved with irrigation at 2 and potassium at 1.
(d) From Table 6.7d and Fig. 6.10d, with a fixed phosphorus fertilizer application level, yield increases rapidly with nitrogen fertilizer application, peaking at a nitrogen level of 1, before gradually decreasing. With a constant nitrogen

Table 6.7 (a) Interaction of irrigation amount and nitrogen application on yield

Program		N (kg/667 m^2)				
		−2	−1	0	1	2
Irrigation amount (m^3/667 m^2)	−2	1190.31	1241.04	1258.09	1241.47	1191.18
	−1	1176.42	1246.32	1282.54	1285.10	1253.98
	0	1162.53	1251.60	1307.00	1328.72	1316.78
	1	1148.64	1256.88	1331.45	1372.35	1379.58
	2	1134.75	1262.16	1355.91	1415.98	1442.38

(b) Interaction of irrigation amount and phosphorus fertilizer application on yield

项目		P_2O_5 (kg/667 m^2)				
		−2	−1	0	1	2
Irrigation amount (m^3/667 m^2)	−2	1076.95	1199.90	1258.09	1251.51	1180.18
	−1	1101.40	1224.35	1282.54	1275.97	1204.63
	0	1125.86	1248.81	1307.00	1300.42	1229.09
	1	1101.40	1224.35	1282.54	1275.97	1204.63
	2	1076.95	1199.90	1258.09	1251.51	1180.18

(c) Interaction of irrigation amount and potassium fertilizer application on yield

项目		K_2O (kg/667 m^2)				
		−2	−1	0	1	2
irrigation amount (m^3/667 m^2)	−2	1204.39	1256.26	1258.09	1209.87	1111.60
	−1	1183.10	1257.84	1282.54	1257.19	1181.79
	0	1161.81	1259.43	1307.00	1304.52	1251.99
	1	1140.52	1261.01	1331.45	1351.85	1322.19
	2	1119.23	1262.59	1355.91	1399.17	1392.39

(d) Interaction of nitrogen fertilizer application and phosphorus fertilizer application on yield

项目		P_2O_5 (kg/667 m^2)				
		−2	−1	0	1	2
N(kg/667 m^2)	−2	981.39	1104.34	1162.53	1155.96	1084.62
	−1	1070.46	1193.41	1251.60	1245.02	1173.69
	0	1125.86	1248.81	1307.00	1300.42	1229.09
	1	1147.58	1270.53	1328.72	1322.15	1250.81
	2	1135.64	1258.59	1316.78	1310.21	1238.87

(e) Interaction of nitrogen fertilizer application and potassium fertilizer application on yield

项目		K_2O (kg/667 m^2)				
		−2	−1	0	1	2
N(kg/667 m^2)	−2	938.99	1075.78	1162.53	1199.23	1185.88
	−1	1067.23	1184.44	1251.60	1268.71	1235.77
	0	1161.81	1259.43	1307.00	1304.52	1251.99

(continued)

Table 6.7 (continued)

(e) Interaction of nitrogen fertilizer application and potassium fertilizer application on yield

项目		K_2O (kg/667 m^2)				
		−2	−1	0	1	2
	1	1222.71	1300.74	1328.72	1306.66	1234.55
	2	1249.94	1308.38	1316.78	1275.13	1183.43

(f) Interaction of phosphorus fertilizer application and potassium fertilizer application on yield

项目		K_2O (kg/667 m^2)				
		−2	−1	0	1	2
P_2O_5 (kg/667 m^2)	−2	980.66	1078.28	1125.86	1123.38	1070.85
	−1	1103.62	1201.24	1248.81	1246.33	1193.80
	0	1161.81	1259.43	1307.00	1304.52	1251.99
	1	1155.23	1252.85	1300.42	1297.95	1245.42
	2	1083.89	1181.51	1229.09	1226.61	1174.08

level, yield also rises and then declines with increasing phosphorus application, reaching its highest when phosphorus is at 0. The maximum yield in the nitrogen-phosphorus interaction is achieved with nitrogen at 1 and phosphorus at 0.

(e) From Table 6.7e and Fig. 6.10e, when potassium fertilizer levels are between −2 and −1, there is a positive correlation between nitrogen fertilizer application and strawberry yield. However, at potassium levels between 0 and 2, the relationship between nitrogen application and yield first increases and then decreases. With a fixed nitrogen level, yield increases rapidly with potassium fertilizer application until reaching a peak at a potassium level of 1, then declines. The optimal yield in the nitrogen-potassium interaction occurs with nitrogen at 1 and potassium at 0.

(f) From Table 6.7f and Fig. 6.10f, when both nitrogen and potassium fertilizer levels are fixed, their effects on yield show a trend of first increasing and then decreasing. The maximum yield is obtained when phosphorus and potassium levels are both at 0.

Based on the established irrigation, nitrogen, phosphorus, and potassium coupling model, frequency analysis using DPS software indicates that among the 140 scenarios with yields exceeding 1239.45 kg/667 m^2, the optimal coupling scheme for irrigation, nitrogen, phosphorus, and potassium is as follows: irrigation amount of 50.95–52.51 m^3/667 m^2, nitrogen fertilizer application of 10.51–11.84 kg/667 m^2, phosphorus fertilizer application of 6.87–7.83 kg/667 m^2, and potassium fertilizer application of 9.55 to 10.97 kg/667 m^2. The optimal fertilizer ratio is N:P_2O_5:K_2O = 1:0.66:0.92. The highest yield is achieved with the following factor levels: irrigation amount $x_1 = 2$, nitrogen fertilizer $x_2 = 2$, phosphorus fertilizer $x_3 = 0$, and potassium fertilizer $x_4 = 1$. This corresponds to an irrigation amount of 57.118 m^3/667 m^2, nitrogen fertilizer application of 16.093 kg/667 m^2, phosphorus fertilizer application of 5.930 kg/667

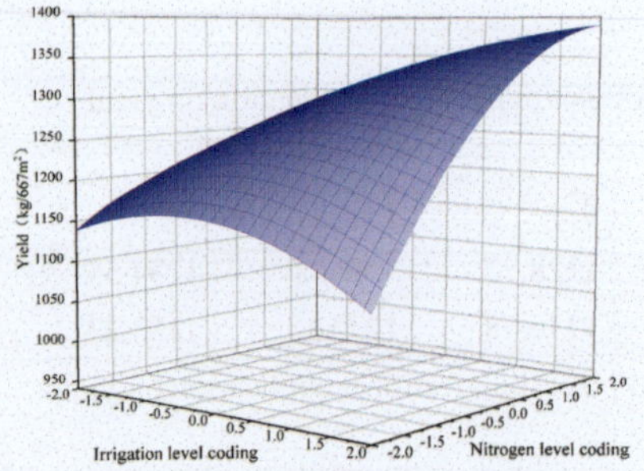

a: Interaction between Irrigation Amount and Nitrogen Fertilizer

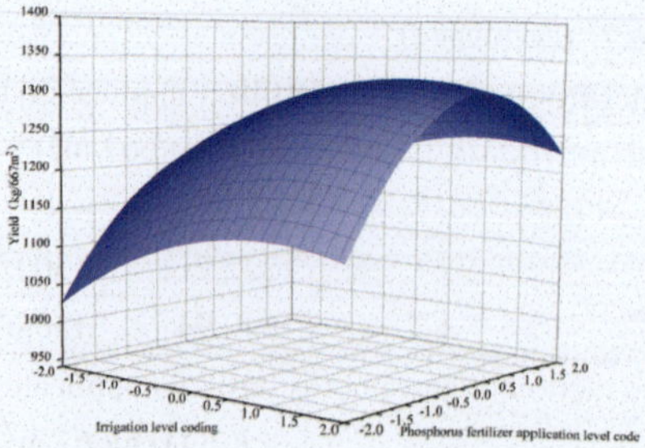

b: Interaction between Irrigation Amount and Phosphorus Fertilizer

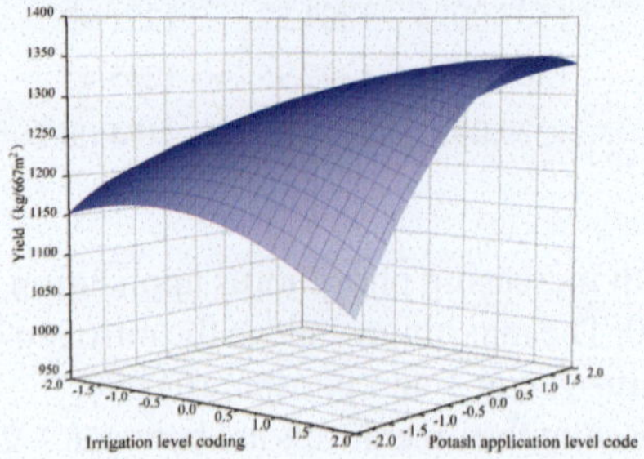

c: Interaction between Irrigation Amount and Potassium Fertilizer

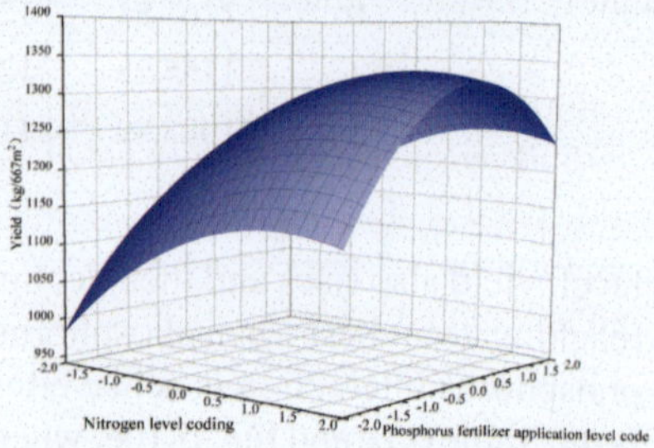

d: Interaction between Nitrogen Fertilizer and Phosphorus Fertilizer

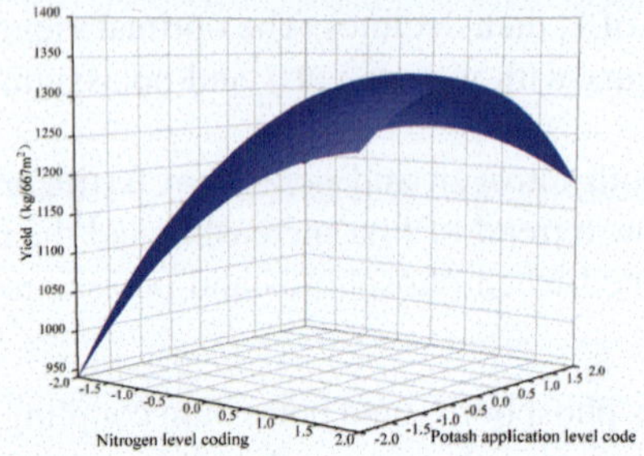

e: Interaction between Nitrogen Fertilizer and Potassium Fertilizer

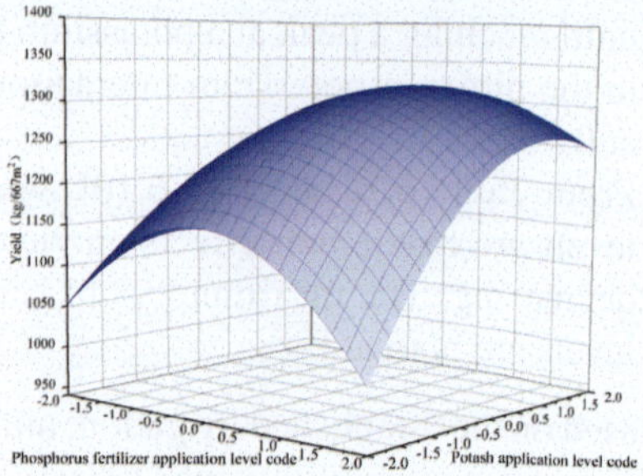

f: Interaction between Phosphorus Fertilizer and Potassium Fertilizer

Fig. 6.10 Response surfaces of interactions among factors

m^2, and potassium fertilizer application of 12.069 kg/667 m^2, resulting in a yield of 1446.47 kg/667 m^2.

6.3.6 *Relationship Model Between Nitrogen, Phosphorus, Potassium, and Electrical Conductivity*

6.3.7 *Application of Electrical Conductivity in Water and Fertilizer Management*

Electrical Conductivity (EC) measures the ability of a substance to conduct an electric current. In agriculture and horticulture, EC is used to assess the concentration of ions in a water solution. Specifically, EC gauges how well dissolved ions—such as sodium, potassium, calcium, magnesium, chloride, and sulfate—conduct electricity. It is expressed in Siemens per meter (S/m) or milliSiemens per centimeter (mS/cm).

In the context of water and fertilizer management, EC reflects the concentration of Total Dissolved Solids (TDS) in a solution, which includes nutrients, fertilizer components, and other dissolved substances. A higher EC value indicates a higher concentration of dissolved ions, which corresponds to a higher concentration of fertilizer. By measuring the EC of irrigation water or fertilizer solutions, growers can accurately determine the concentration of fertilizers and adjust their application to ensure that strawberries receive the optimal amount of nutrients, avoiding both excess and deficiency.

High EC values often signal elevated salt concentrations in the soil or water. Excessive salt can damage strawberry plants by impairing the roots' ability to absorb water and nutrients, potentially leading to reduced yield and quality. Monitoring EC levels helps mitigate salt damage by enabling adjustments in fertilizer and irrigation practices to control salt levels effectively.

Strawberries have varying water and nutrient requirements at different growth stages. EC measurements can guide growers in tailoring water and fertilizer applications to these needs. For instance, during the early growth stage, a lower EC fertilizer solution can support root development, while during the fruit enlargement stage, a higher concentration may be necessary to promote fruit growth. Effective EC management thus plays a crucial role in enhancing strawberry growth and fruit quality. Strawberries have varying water and nutrient requirements at different growth stages. EC measurements can guide growers in tailoring water and fertilizer applications to these needs. For instance, during the early growth stage, a lower EC fertilizer solution can support root development, while during the fruit enlargement stage, a higher concentration may be necessary to promote fruit growth. Effective EC management thus plays a crucial role in enhancing strawberry growth and fruit quality.

6.3.8 Experimental Study: Establishing the Relationship Model Between Nitrogen, Phosphorus, Potassium, and Electrical Conductivity

Integrated water and fertilizer equipment primarily uses Electrical Conductivity (EC) as a decision-making indicator. In horticultural planting, however, the concentration of the fertilizer solution is typically used to guide application. Despite the optimization of irrigation and fertilization amounts in previous studies, the relationship between fertilizer solution concentration and EC remains unclear. To address the integration gap between agronomy and agricultural machinery, this study employs a three-factor, five-level orthogonal experiment to examine how EC varies with different effective nutrient concentrations. This research aims to develop a model correlating nitrogen, phosphorus, and potassium concentrations with EC, facilitating the conversion of agronomic fertilizer concentrations into machine-readable EC levels, thus enhancing integrated water and fertilizer application.

6.3.8.1 Experimental Materials and Scheme

Experimental Fertilizers: The fertilizers used are urea (N: 46%), superphosphate (P_2O_5: 16%), and potassium sulfate (K_2O: 52%).

Water Source: Local groundwater is used for the experiment.

Preparation of Solutions: Fertilizers are calculated based on their effective content and dissolved in water to create solutions of varying concentrations.

Experimental Design: A three-factor, five-level orthogonal experiment is conducted with effective concentrations of nitrogen, phosphorus, and potassium fertilizers as the experimental factors. Each factor is set at five levels based on effective concentration:

Nitrogen (N) Concentrations: 0.0‰, 0.2‰, 0.4‰, 0.6‰, 0.8‰

Phosphorus (P_2O_5) Concentrations: 0.0‰, 0.1‰, 0.2‰, 0.3‰, 0.4‰

Potassium (K_2O) Concentrations: 0.0‰, 0.1‰, 0.2‰, 0.3‰, 0.4‰

The natural factor levels are encoded as shown in Table 6.8, and the design structure matrix is presented in Table 6.9.

Table 6.8 Natural factors level coding table

Factor levels	x_1 (N: ‰)	x_2 (P_2O_5: ‰)	X_3 (K_2O: ‰)
1	0.0‰	0.0‰	0.0‰
2	0.2‰	0.1‰	0.2‰
3	0.4‰	0.2‰	0.4‰
4	0.6‰	0.3‰	0.6‰
5	0.8‰	0.4‰	0.8‰

Table 6.9 Structure matrix for the three-factor five-level orthogonal experimental design

Treatment	Factor		
	X_1 (N: ‰)	X_2 (P_2O_5: ‰)	X_3 (P_2O_5: ‰)
T1	1 (0.0‰)	1 (0.0‰)	1 (0.0‰)
T2	1 (0.0‰)	2 (0.1‰)	3 (0.4‰)
T3	1 (0.0‰)	3 (0.2‰)	5 (0.8‰)
T4	1 (0.0‰)	4 (0.3‰)	2 (0.2‰)
T5	1 (0.0‰)	5 (0.4‰)	4 (0.6‰)
T6	2 (0.2‰)	1 (0.0‰)	5 (0.8‰)
T7	2 (0.2‰)	2 (0.1‰)	2 (0.2‰)
T8	2 (0.2‰)	3 (0.2‰)	4 (0.6‰)
T9	2 (0.2‰)	4 (0.3‰)	1 (0.0‰)
T10	2 (0.2‰)	5 (0.4‰)	3 (0.4‰)
T11	3 (0.4‰)	1 (0.0‰)	4 (0.6‰)
T12	3 (0.4‰)	2 (0.1‰)	1 (0.0‰)
T13	3 (0.4‰)	3 (0.2‰)	3 (0.4‰)
T14	3 (0.4‰)	4 (0.3‰)	5 (0.8‰)
T15	3 (0.4‰)	5 (0.4‰)	2 (0.2‰)
T16	4 (0.6‰)	1 (0.0‰)	3 (0.4‰)
T17	4 (0.6‰)	2 (0.1‰)	5 (0.8‰)
T18	4 (0.6‰)	3 (0.2‰)	2 (0.2‰)
T19	4 (0.6‰)	4 (0.3‰)	4 (0.6‰)
T20	4 (0.6‰)	5 (0.4‰)	1 (0.0‰)
T21	5 (0.8‰)	1 (0.0‰)	2 (0.2‰)
T22	5 (0.8‰)	2 (0.1‰)	4 (0.6‰)
T23	5 (0.8‰)	3 (0.2‰)	1 (0.0‰)
T24	5 (0.8‰)	4 (0.3‰)	3 (0.4‰)
T25	5 (0.8‰)	5 (0.4‰)	5 (0.8‰)

6.3.8.2 Experimental Results and Analysis

Electrical conductivity sensors were used to measure the different treatment scenarios. The results are shown in Table 6.10.

Based on the results of the orthogonal experiment, the electrical conductivity (EC) of the local groundwater was measured at 1.43 mS/cm. This level of EC is generally considered unsuitable for optimal strawberry growth. However, observations from this study revealed that the strawberry plants grew normally despite this conductivity condition. This anomaly suggests that the strawberry seedlings used in the study, having been locally bred over many years, may have developed a certain level of adaptability to saline-alkaline water conditions.

Table 6.10 Electrical conductivity results for different treatments

Treatment	Electrical conductivity (mS/cm)	Treatment	Electrical conductivity (mS/cm)	Treatment	Electrical conductivity (mS/cm)
T1	1.43	T10	2.72	T18	2.99
T2	2.13	T11	3.40	T19	3.73
T3	2.88	T12	2.22	T20	2.62
T4	1.97	T13	3.00	T21	3.42
T5	2.68	T14	3.57	T22	4.01
T6	3.75	T15	2.46	T23	3.04
T7	2.34	T16	3.53	T24	3.49
T8	3.05	T17	4.01	T25	4.58
T9	1.91	–	–	–	–

Following these observations, a second-order polynomial regression analysis was performed using DPS software. The resulting regression model is provided in Eq. (6.19).

$$Y = 1.54 + 2.07x_1 - 1.05x_2 + 1.90x_3 - 0.18x_1^2 + 2.78x_2^2 + 0.15x_3^2 - 0.30x_1x_2 - 0.16x_1x_3 - 0.33x_2x_3 \quad (6.19)$$

The regression equation after fitting has a correlation coefficient $R^2 = 0.98$, indicating that the fitted equation effectively reflects the relationship between different nutrient concentrations and electrical conductivity.

Based on the guidelines provided in Chap. 2 of this paper, which outline the optimal fertilization and irrigation amounts for facility-grown strawberries, combined with local management practices and the nutrient requirements at different growth stages, the following approach is recommended:

1. Leaf Expansion Stage: For a single fertilization event during the leaf expansion stage, use the application amount specified as the standard.
2. Flowering and Blooming Stages: During these stages, increase the fertilization amount to 1.5 times that of the leaf expansion stage.
3. Fruit Swelling to Maturity Stages: For these later stages, the fertilization amount should be twice that of the leaf expansion stage.

The single fertilization and irrigation amounts are determined based on the average daily irrigation requirement throughout the entire growth cycle. This approach allows for the estimation of optimal single fertilization amounts and the corresponding electrical conductivities for various growth periods. These details are summarized in Table 6.11.

Table 6.11 Optimal fertilization amounts and corresponding electrical conductivities for different periods

Growth stage	N (kg/667 m^2)	P_2O_5 (kg/667 m^2)	K_2O (kg/667 m^2)	电导率 (mS/cm)
Leaf expansion stage	0.050–0.056	0.033–0.037	0.045–0.052	1.915–1.954
Budding to full bloom stage	0.075–0.085	0.049–0.056	0.068–0.078	2.108–2.169
Fruit swelling to maturity stage	0.100–0.113	0.065–0.075	0.091–0.104	2.306–2.389

6.3.9 Application of the Electrical Conductivity Model

In the design of the water-fertilizer coupling experiments, various schemes were evaluated to refine the model's accuracy. Ultimately, a second-order general rotational combination experimental design was chosen. To minimize the impact of external environmental factors on different treatments, 30 experimental groups were spatially distributed and received manual fertilization and irrigation daily. During the experiment, it was noted that the surface layer of the sandy substrate exhibited slight hardening and compaction. This was attributed to the high mineralization of the local water source and the small particle size of the sandy substrate, which led to salt accumulation on the surface due to prolonged fertilization and irrigation. The uniformity of nutrient distribution within the sandy substrate still requires further verification.

The analysis of factor interactions revealed that there are specific thresholds for fertilization and irrigation quantities. Within these thresholds, increasing fertilization and irrigation amounts leads to significant improvements in yield. However, as the thresholds are approached, the effect of each factor on yield gradually diminishes, demonstrating an overall positive correlation. Beyond these thresholds, excessive water and fertilizer application begins to inhibit yield, with increasing irrigation, nitrogen, phosphorus, and potassium amounts causing greater inhibitory effects, thus exhibiting an overall negative correlation. This finding aligns with the research conclusions of Zhang Jing et al.

By employing the second-order general rotational combination experimental design, the study determined the optimal fertilization and irrigation amounts for facility strawberries in southern Xinjiang. The results indicate that the impact on yield is ranked as follows: nitrogen application amount > phosphorus application amount > irrigation amount > potassium application amount. For treatments yielding more than 1,239.45 kg/667 m^2, the optimal ranges are as follows: irrigation amount between 50.95 and 52.51 m^3/667 m^2, nitrogen application between 10.51 and 11.84 kg/667 m^2, phosphorus application between 6.87 and 7.83 kg/667 m^2, and potassium application between 9.55 and 10.97 kg/667 m^2. The optimal nutrient ratio for these conditions is $N:P_2O_5:K_2O = 1:0.66:0.92$.

Additionally, this study establishes a model correlating different fertilization concentrations with electrical conductivity, which provides a basis for future fertilization decisions in integrated water and fertilizer systems. The optimal electrical conductivities for strawberries at different growth stages are as follows: 1.93 mS/cm during the leaf expansion period, 2.13 mS/cm during the bud flowering period, and 2.34 mS/cm during the fruit maturity and enlargement period.

6.3.10 Research Conclusions on Facility Strawberries in Desert Gobi

In the study of facility strawberries in southern Xinjiang, an optimized evapotranspiration model was first established to better understand the water consumption of strawberries in controlled environments. Using the Penman–Monteith evapotranspiration model, theoretical evapotranspiration values were compared with actual measurements from continuous environmental data collected in the greenhouse. The optimization process resulted in a substantial improvement: the maximum and minimum errors of the unoptimized model were 2.185 mm and 0.603 mm, respectively, with a mean absolute error (MAD) of 1.279 mm. After optimization, the maximum and minimum errors were reduced to 0.313 mm and 0.007 mm, respectively, with the MAD decreasing to 0.117 mm and the mean square error (MS) at 2.2045. The error compared to actual values was only 6.05%. This optimized model more accurately reflects the water consumption of strawberries in the greenhouse, providing a reliable basis for determining the optimal irrigation amount and developing scientifically-based irrigation plans for facility strawberries in southern Xinjiang.

In the realm of water-fertilizer management, a water-fertilizer coupling model for facility strawberries was developed. This model utilized a second-order general rotational combination experimental design to explore the impact of varying irrigation and fertilization amounts on strawberry yield. The results revealed the following ranking of factors influencing strawberry yield: nitrogen application amount > phosphorus application amount > irrigation amount > potassium application amount. The analysis identified specific thresholds for each factor's application amount. Within these thresholds, yield showed a positive correlation with the application amounts. However, beyond these thresholds, further increases in water and fertilizer application amounts may lead to diminishing returns and potential yield inhibition.

The optimal ranges determined by frequency analysis are as follows: irrigation amount of 50.95 to 52.51 $m^3/667\ m^2$, nitrogen application of 10.51 to 11.84 kg/667 m^2, phosphorus application of 6.87 to 7.83 kg/667 m^2, and potassium application of 9.55 to 10.97 kg/667 m^2. The best nutrient ratio identified is $N:P_2O_5:K_2O = 1:0.66:0.92$. The maximum yield of 1446.47 kg/667 m^2 was achieved with an irrigation amount of 57.118 $m^3/667\ m^2$, nitrogen application of 16.093 kg/667 m^2, phosphorus application of 5.930 kg/667 m^2, and potassium application of 12.069 kg/667 m^2.

Additionally, by establishing the relationship model between nitrogen, phosphorus, potassium, and electrical conductivity, as well as the water-fertilizer management model, the optimal electrical conductivities for strawberries at different growth stages were determined: 1.93 mS/cm during leaf expansion, 2.13 mS/cm during bud flowering, and 2.34 mS/cm during fruit maturity and enlargement. These findings offer significant technical support for the efficient cultivation of facility strawberries in southern Xinjiang.

Chapter 7
Design and Optimization of Intelligent Water and Fertilizer Integration System for Desert and Gobi Facilities

Wenhao Xia, Can Hu, Ji Shi, Shufeng Zhuang, Xiaoqin Lian, and Menghao Shi

As global climate change intensifies and resource scarcity grows, agricultural production encounters unprecedented challenges. In this context, achieving efficient agricultural practices in harsh environments, particularly in resource-limited regions such as deserts and the Gobi has emerged as a critical concern in agricultural science and technology. The vast Gobi Desert in southern Xinjiang offers significant potential for expanding facility-based agriculture. However, traditional agricultural methods in these regions are hindered by severe water scarcity, extensive soil salinization, and low fertilizer efficiency. Therefore, the development and implementation of innovative technologies and approaches are essential to enhance the sustainability of agricultural production.

Facility agriculture, as a key component of modern agricultural development, has gradually been introduced in desert and Gobi regions due to its ability to control environmental conditions and optimize resource utilization. However, most current water and fertilizer management practices in facility agriculture still rely heavily on

W. Xia · C. Hu (✉) · J. Shi · S. Zhuang · X. Lian
College of Mechanical and Electrical Engineering, Tarim University, Alar 843300, China
e-mail: hucanboy1@qq.com

Modern Agricultural Engineering Key Laboratory at Universities of Education Department of Xinjiang Uygur Autonomous Region, Alar 843300, China

Xinjiang Production and Construction Corps (XPCC) Key Laboratory of Utilization and Equipment of Special Agricultural and Forestry Products in Southern Xinjiang, Alar 843300, China

M. Shi
Key Laboratory of Smart Agriculture System Integration, Ministry of Education, Beijing 100083, China

Key Laboratory of Agricultural Information Acquisition Technology, Ministry of Agriculture and Rural Affairs, Beijing 100083, China

C. Hu et al. (eds.), *New Technologies Applied in Gobi Desert Facility Agriculture*, Smart Agriculture 11, https://doi.org/10.1007/978-981-96-1031-0_7

manual operations and experiential knowledge, leading to low efficiency in resource use and contributing to waste and environmental pollution. Additionally, the unique soil characteristics of desert and Gobi regions significantly restrict plant root growth, further impacting crop yield and quality.

To address these challenges, the integration of intelligent water and fertilizer systems is emerging as a promising solution. By utilizing real-time monitoring of environmental parameters combined with precise control algorithms, these systems can automatically adjust water and fertilizer applications based on the actual needs of crops, thereby achieving precise and efficient resource utilization. In desert and Gobi regions, in particular, intelligent water and fertilizer integration systems not only significantly enhance the efficiency of water and fertilizer use but also improve crop growing conditions, thereby increasing both yield and quality. Thus, exploring and optimizing the design of intelligent water and fertilizer integration systems tailored to the specific needs of desert regions is a critical area of research. By constructing a water and fertilizer deployment model tailored to local climate and soil conditions, and integrating modern Internet of Things technology with automatic control systems, this study designs an intelligent water and fertilizer integration system based on PLC control. The system underwent experimental validation and performance optimization. The results demonstrate that the system significantly improves the efficiency of water and fertilizer use, reduces resource waste, and enhances both crop yield and quality. This work provides a feasible and efficient solution for water and fertilizer management in facility agriculture, particularly in desert and Gobi regions.

The research presented in this paper not only offers technical support for the advancement of facility-based agriculture in desert and Gobi regions but also serves as a valuable reference for agricultural production under similar climatic conditions globally. The successful application of this research is expected to have a profound impact on the future of agriculture, driving it toward greater intelligence, efficiency, and sustainability.

7.1 Background and Significance of Water and Fertilizer Integration in Facility Agriculture

7.1.1 Current Research Status of Water and Fertilizer Integration Technology

As early as 2,800 years ago in ancient Greece, the city-state of Athens pioneered the use of urban wastewater for agricultural irrigation, representing an early exploration of water and fertilizer integration technology. However, due to the technological constraints of that era, the advancement of this technology progressed relatively slowly. In 1925, with the advent of greenhouse technology, nutrient solutions began to replace soil as a medium for cultivation. This practice is recognized as an early prototype of modern water and fertilizer integration technology. The advancement of water and fertilizer integration technology is closely linked to the progress in drip irrigation technology. In 1933, Robey O.E. first attempted to use a canvas pipe system for underground drip irrigation. However, due to material limitations that led to significant water loss, this method was not widely adopted. It was not until the 1950s, with the rapid development of the plastics industry, that plastic-based drip irrigation and micro-sprinkler systems began to gain popularity. In 1960, Israeli engineer Simcha B. developed the world's first drip irrigation system, characterized by its simple structure and the use of holes or micro-tubes instead of complex watering devices. This system was primarily employed for irrigating small orchards, flower gardens, and vegetable gardens. However, due to incomplete design, early drip irrigation systems were prone to clogging problems during fertilization and irrigation. To address these challenges, researchers focused on optimizing runner structures and investigating the causes of drip head clogging. The development of the Turbolent™ runner structure by NETAFIM in Israel significantly enhanced drip head performance. By the mid-1970s, drip irrigation and fertigation had been successfully adopted for field crop management in countries such as Australia, Israel, Mexico, New Zealand, and South Africa.

In the early stages, water and fertilizer were mixed using mixing tanks. However, this method often resulted in poor uniformity in irrigation fertilization. To address this issue, researchers developed the Venturi fertilizer applicator and water pressure-driven fertilizer injector, which significantly enhanced the uniformity of fertilizer application. With the advent of the information age, the integration of various sensors and microprocessors has further advanced the precision and intelligence of fertilizer application, leading to gradual improvements in the overall framework of water and fertilizer integration technology.

In China, research on water and fertilizer integration technology began relatively late. Since the introduction of drip irrigation equipment in the 1970s, the technology has undergone several phases, including initial exploration, pilot promotion, and widespread adoption. Early on, scholars proposed dissolving fertilizer in water for drip irrigation to address the challenge of combining fertilizer application with the drip irrigation method. This approach represents the initial phase of water and fertilizer integration research in China. In 1980, China successfully developed the first generation of complete drip irrigation systems, marking the beginning of the integration of water and fertilizer technology with sprinkler and drip irrigation methods. This development was followed by extensive experimentation and demonstration, leading to gradual nationwide adoption. In 1988, Zhiang Xu systematically reviewed several major fertilizer application methods of the time, establishing a foundation for contemporary water and fertilizer integration technology. In 1996, the Xinjiang Production and Construction Corps applied water and fertilizer integration technology to large-scale cotton cultivation for the first time. Around 2000, the technology entered a phase of rapid promotion, becoming widely used in facility agriculture and significantly enhancing crop yields. With the gradual establishment of China's water-soluble fertilizer industry and improvements in supporting products, a preliminary industrial chain for water and fertilizer integration equipment began to take shape.

Driven by a series of water conservation and fertilizer reduction policies, water and fertilizer integration technology has increasingly become prevalent nationwide.

7.1.2 Overview of Research on Water and Fertilizer Integration in Desert Gobi Facility Agriculture

Deserts and the Gobi region cover extensive areas in northwestern China, particularly in Xinjiang, where desertification severely restricts the development of traditional agriculture. However, the unique geographic and climatic conditions of these regions—such as abundant sunshine and a dry climate—offer potential advantages for facility-based agriculture. By artificially controlling environmental conditions, facility-based agriculture can achieve efficient production even in harsh natural environments. Therefore, effectively leveraging the resources of deserts and the Gobi region for the development of facility agriculture has become a significant tropic in modern agricultural advancement.

The main challenges to the development of facility-based agriculture in desert and Gobi regions are the efficient use of water resources and the precise fertilization management of crops. Traditional irrigation and fertilization methods often suffer from low water and fertilizer utilization and serious resource wastage. In these areas, water resources are extremely valuable, and over-irrigation not only wastes water but also exacerbates the problem of soil salinization, which affects the normal growth of crops. In addition, traditional fertilizer application methods make it difficult to accurately control the amount of fertilizer used, resulting in inefficient use of fertilizer, increasing the cost of agricultural production and potentially polluting the environment.

To address these challenges, intelligent water and fertilizer integration systems are emerging as an ideal solution. By integrating Internet of Things (IoT) technology, sensor technology, and automatic control systems, these systems can monitor key parameters in the crop growing environment, such as soil moisture, conductivity, temperature, and humidity in real time. They can then automatically adjust the application of water and fertilizer based on the actual needs of the crops, achieving precise irrigation and fertilization. This approach not only significantly improves the efficiency of water and fertilizer use and reduces agricultural production costs but also enhances the growing environment for crops, leading to increased yield and improved quality.

In this context, this paper investigates the design and optimization of an intelligent water and fertilizer integration system tailored for desert and Gobi regions, with a focus on facility agriculture in southern Xinjiang. The research encompasses several key areas: developing a water and fertilizer deployment model adapted to the climatic conditions of the desert and Gobi regions, designing an intelligent water and fertilizer integration system based on PLC control, and validating the system's performance while optimizing control strategies to achieve precise and efficient resource utilization. The ultimate aim of this research is to provide technical support for the advancement of facility-based agriculture in these regions and to promote the modernization and sustainable development of agriculture in desert and Gobi areas.

7.1.3 Current Issues with Domestic Water and Fertilizer Integration Equipment

A review of domestic and international literature reveals that, although extensive research has been conducted on the water and fertilizer requirements of various crops, most studies focus primarily on field crops. Due to regional differences in climate, management practices, and crop varieties, the existing water and fertilizer coupling models are challenging to apply to strawberry cultivation in southern border facilities. Current research on water and fertilizer integration equipment provides valuable theoretical and experimental insights but lacks large-scale practical applications. Many researchers have focused on algorithmic improvements to enhance fertilizer control accuracy, aiming for more precise fertilization. Intelligent water and fertilizer integration systems can adjust the conductivity of the fertilizer for accurate application. However, the absence of a well-established water and fertilizer deployment model leaves growers uncertain about the appropriate amount of fertilizer to apply at different growth stages. The integration of agro-mechanical and agronomic approaches remains insufficient, preventing the realization of truly need-based fertilization and indicating that the level of intelligence in current water and fertilizer integration equipment is still inadequate.

(1) Absence of a water and fertilizer deployment model for facility strawberries in the southern Xinjiang region: Although extensive research has been conducted on the water and fertilizer requirements of various crops, the existing water and fertilizer coupling models are challenging to apply to facility strawberry cultivation in South Xinjiang. This difficulty arises from the diverse climatic conditions, management practices, and crop varieties across different regions. Consequently, the specific fertilizer requirements for facility-grown strawberries in South Xinjiang during various growth stages remain unclear.

(2) Insufficient integration between water and fertilizer equipment and agronomic practices. Currently, most research efforts focus on refining algorithms to enhance the precision of fertilization and irrigation. Although existing control algorithms have largely achieved accurate fertilization, further improvements yield diminishing returns in overall system performance. Additionally, there is limited research on integrating crop-specific fertilizer models with water and fertilizer equipment, leading to continued reliance on subjective human judgment to determine crop fertilizer requirements. This highlights the need for stronger integration between agro-mechanical systems and agronomic practices.

7.2 Functional Analysis of Water and Fertilizer Integration System

7.2.1 *Analysis of System Functions*

Based on the specific water and fertilizer requirements for strawberries cultivated in yellow sand substrate containers in South Xinjiang, as well as the prevailing management practices, the water and fertilizer integration system is composed of several key components, each fulfilling a distinct role and coordinating with others to achieve efficient irrigation and fertilization. The system can be functionally divided into several parts: water source engineering, the primary hub, fertilizer application device, field pipe network, and irrigation units. The water source engineering provides a stable supply of irrigation water, typically drawn from rivers, lakes, or groundwater. The primary hub pressurizes this water via a supply pump, filters it, and then delivers the nutrient-enriched solution to the field pipe network. As the system's core, the primary hub consists of four essential components: a filtration device, a fertilizer application system, a pressure and flow monitoring and protection system, and an automation control system. These elements collectively drive, measure, and regulate the system's operations, forming the central control unit for the entire setup.

The fertilizer application system is a crucial component of the water and fertilizer integration system. It is typically installed at the front end of the primary hub or irrigator, depending on its characteristics and the specific requirements of irrigation and fertilization. This system is responsible for accurately injecting fertilizers into the water flow to achieve precise fertilization. The monitoring and control system plays a critical role in ensuring the safe operation of the overall system. It regulates the process of fertilizer application and irrigation, ensuring that the water-to-fertilizer ratio and the amount of fertilizer applied align with the crop's specific needs.

The field piping network is a critical component responsible for transporting the liquid water and fertilizer mixture to the crop roots. It primarily consists of trunk, branch, and capillary pipes, which are strategically arranged to ensure that the pressurized water flow is effectively distributed to each irrigation unit and irrigator as required. As the terminal equipment of the system, the design and water discharge method of the irrigator directly influence irrigation efficiency. Irrigators can be categorized into different types, including sprinkler irrigation, micro-sprinkler irrigation, drip irrigation, and seepage irrigation, based on their structure and discharge method. Sprinkler irrigation, though effective, is susceptible to wind, has high evaporation losses, and may promote weed growth and pest issues, making it more suitable for large field irrigation in humid climates. Seepage irrigation is less commonly used in practice due to the ease of clogging in drip heads. Currently, drip irrigation is the most widely adopted method within water and fertilizer integration systems, particularly in facility-based agricultural production and the cultivation of dry-farm cash crops in extensive fields. The arrangement of the field pipe network and irrigators must be tailored to the terrain, planting scheme, and natural environment to achieve optimal irrigation efficiency. A detailed schematic of the water and fertilizer integration system is provided below. The equipment designed in this study meets the following requirements.

(1) Uniform mixing and control of water and fertilizer. The water and fertilizer integration equipment must include a mother liquor tank and a fertilizer mixing tank. Using a pre-established water and fertilizer blending model, along with optimal fertilization conductivity, the system transfers fertilizer from the mother liquor tank into the mixing tank. By employing a control strategy, the system precisely regulates the input of the mother liquor, ensuring accurate control and adjustment of the water and fertilizer conductivity.
(2) Automated fertilization and irrigation. The system integrates Internet of Things technology, automatic control technology, and water and fertilizer integration technology to enable automated fertilization and irrigation. By collecting data from environmental sensors, the system processes this information through internal programs within the processor to make real-time irrigation decisions. Based on the actual planting and management practices, the system also autonomously determines the appropriate timing and amount of fertilization.
(3) Real-Time monitoring of the water and fertilizer integration system. The water and fertilizer integration equipment includes a human–computer interaction (HCI) interface, connected to a lower-level computer, which provides real-time updates on the operation of environmental equipment and surrounding conditions. The HCI interface not only displays operational parameters but also offers manual and automatic control functions based on the system's production needs. The design of the HCI page should be simple and intuitive, ensuring ease of use and effective interaction.

7.2.2 *Control System Architecture*

Based on the functional requirements of the water and fertilizer integration equipment and existing research on water and fertilizer integration technology, an intelligent water and fertilizer integration system tailored for strawberry cultivation in facilities has been designed. This system comprises three main components: the information acquisition system, the control and execution system, and the human–computer interaction interface. The information acquisition system includes soil temperature and humidity sensors, soil electrical conductivity (EC) sensors, fertilizer barrel EC sensors, fertilizer barrel pH sensors, and liquid level sensors. These sensors are connected to the lower-level computer via wired connections to collect and transmit environmental data and equipment operation information. This data supports the optimization of the water and fertilizer deployment model and the control and execution system. The control and execution system is the core of the water and fertilizer integration equipment. It consists of a PLC controller, fertilizer suction solenoid valves, water replenishment solenoid valves, and fertilization and irrigation pumps. This system processes information obtained from the acquisition system and, using control

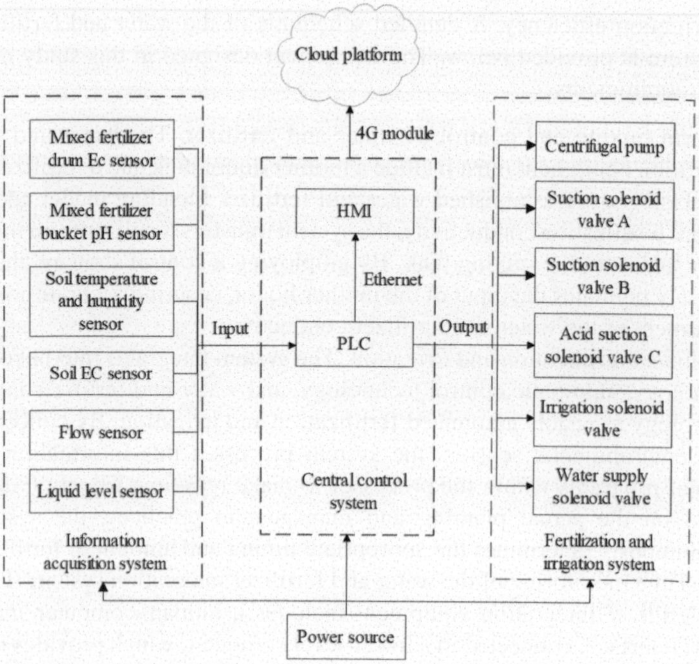

Fig. 7.1 Control architecture sketch

and optimization algorithms, generates execution commands for the solenoid valves. By regulating the opening and closing of these valves, the system precisely adjusts the concentration of water and fertilizer. The human–computer interaction interface facilitates communication with the PLC via Ethernet. The interface design should be intuitive and straightforward, allowing users to easily modify control parameters and monitor equipment status. The control architecture is illustrated in Fig. 7.1.

7.3 Hardware Selection and Design of Water and Fertilizer Integration Equipment

7.3.1 Key Components Selection

7.3.1.1 PLC Selection

Given the high temperature and high humidity environment of daylight greenhouses, ensuring the stability of the control system is crucial. To address this, the Siemens S7-200 SMART CPU ST40 PLC has been selected for the system processor. This PLC model, known for its transistor outputs, is well-suited for high-frequency

Table 7.1 S7-200 SMART PLC and module performance parameters

Type	Parameter name	Parameter content
S7-200 SMART CPU ST40	Supply voltage	24 V DC
	Output method	Transistor output
	Rated output current	300 mA
	Input port	24 digital input ports
	Output port	16 digital output ports
	Communication port	Ethernet, RS485
	Expansion module	Maximum 6
S7-200 SMART EM AE04	Supply voltage	24 V DC
	Number of analog inputs	4
	Input range	Voltage: ±10 V, ±5 V, ±2.5 V
		Current: 0–20 mA, 4–20 mA

control due to its fast response capabilities. It features 24 digital input ports and 16 digital output ports, accommodating the input and output requirements of the entire control system. Additionally, it includes Ethernet and RS485 communication ports, facilitating convenient networking.

Since the Siemens S7-200 SMART CPU ST40 PLC lacks an analog input module for processing analog signals from sensors, an additional analog input module is required. The Siemens S7-200 SMART EM AE04 analog input module has been selected for this purpose. This module can be directly installed as an expansion module for the PLC and includes four analog input channels. Its performance parameters are detailed in Table 7.1.

7.3.1.2 Sensor Selection

The information acquisition system encompasses sensors for substrate temperature and humidity, electrical conductivity (EC), pH, and liquid level in the fertilizer mixing barrel. Specifically, the soil information acquisition sensor selected is the SMET-3A, a three-in-one device that measures humidity, temperature, and EC. This sensor communicates via RS485 and operates within a voltage range of 5–24 V DC, allowing it to provide real-time data on substrate moisture, temperature, and conductivity. Performance parameters for this sensor are detailed in Table 7.2.

The EC and pH sensors in the mixing barrel are critical components of the information acquisition system, ensuring the accuracy of data collection. For conductivity measurement, the Jishen Tongchuang EC-510 model sensor has been selected. This sensor is known for its high accuracy and sensitive response, and it transmits signals via RS485. For pH measurement, the SIN-PH-7002 model sensor from Sinomeasure has been chosen. This sensor offers a measurement range and accuracy that meets

Table 7.2 SMET-3A sensor performance parameters

Parameter name	Conductivity	Moisture content	Temperature
Operating voltage	5–24 V		
Measurement range	0–10 mS/cm	0–100%	−10 to 70 °C
Accurate	2% FS	<3%	<0.3 °C
Communication port	RS485		
Measurement stabilization time	<2 s		
Communication protocols	Modbus		

Table 7.3 EC/pH sensor performance parameters

Type	Parameter name	Parameter content
SIN-PH-7002 pH sensor	Operating voltage	100–240 V
	Measurement range	0–14
	Accurate	±0.1%
	Analog output	Isolated 0/4–20 mA
EC-510 EC sensor	Operating voltage	24 V
	Measurement range	0–200 mS/cm
	Input port	RS485

the design requirements of the system. Performance parameters for these sensors are detailed in Table 7.3.

7.3.1.3 HMI Selection

Due to the intense light, high temperatures, and elevated humidity in solar greenhouses, capacitive screens can be prone to damage in such environments. Therefore, a TFT LCD has been selected for the HMI. Considering both cost and design convenience, the Haiwell C7S-GW has been chosen for the HMI interface. This display integrates a common HMI with an IoT gateway and DTU module, significantly reducing installation space and hardware costs. It is noted for its straightforward programming, low failure rate, and long service life. Performance parameters for the Haiwell C7S-GW are detailed in Table 7.4.

7.3.1.4 Power Switch

The control system includes various electrical components, with the conductivity and pH sensors in the fertilizer tank operating on 240 V AC, while the PLC and analog input expansion module require 24 V DC. Therefore, a power switch is necessary to convert the voltage. To balance performance and cost, the EDR-75-24 model power

Table 7.4 Haiwell C7S-GW HMI performance parameters

Parameter name	Parameter content
Interface	RJ45 port, 2 serial ports (232/422/485), 2 USB ports
Processing unit	4 GB flash + 512 M RAM
Product size W × H × D	200 × 146 × 37 mm
Operating voltage	24 V DC
Monitor	TFT screen

Table 7.5 EDR-75-24 power switch performance parameters

Parameter name	Parameter content
Input voltage	90–264 V AC/127–370 V DC
Output voltage	24 V
Output voltage accuracy	±1.0%
Output current	3.2 A
Output voltage	76.8 W

switch has been selected. Performance parameters for this power switch are detailed in Table 7.5.

7.3.2 *Structural Design of Water and Fertilizer Integration Equipment*

In current water and fertilizer integration systems, there are two primary methods for mixing fertilizers: mixing within the pipe network or mixing in the fertilizer barrel before distributing the mixture through the field network. The former method often results in poor fertilizer uniformity and cannot promptly adjust fertilizer conductivity, making it suitable primarily for crops that are not sensitive to fertilizer concentration. However, this method is not ideal for strawberry cultivation. To enhance both the uniformity of fertilizer mixing and the precision of control, the water and fertilizer integration equipment employs the method of mixing fertilizer in the fertilizer mixing barrel.

The water and fertilizer integration equipment designed in this paper comprises a mother liquor barrel, an acid liquor barrel, a fertilizer mixing barrel, a Venturi fertilizer applicator, a fertilizer pump, various sensors, and an array of piping components. During irrigation operations, the irrigation solenoid valve is opened, the fertilizer pump remains off, and pressurized water is directly supplied to the field pipeline network. For fertilization operations, the water replenishment solenoid valve is activated, and the fertilizer pump is turned on. Based on the specific requirements of different fertilizer formulations, the corresponding mother liquor barrel solenoid valve opens. The system uses EC and pH sensors on the outlet pipe to continuously

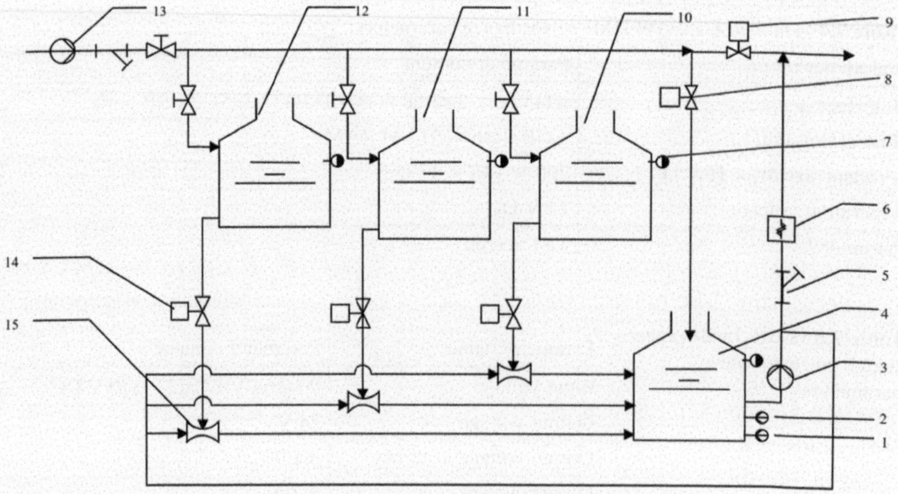

Fig. 7.2 Structure sketch of water and fertilizer integration system

monitor the acidity and conductivity of the mixed fertilizer. The conductivity of the mother liquor is adjusted in real-time by modulating the fertilizer suction solenoid valve. Each barrel is equipped with a liquid level sensor that triggers an alarm and halts water injection if the liquid level exceeds the sensor threshold. Additionally, a flow meter at the system outlet monitors irrigation volume in real-time. The system structure is illustrated in Fig. 7.2.

7.3.3 *Circuit Design*

7.3.3.1 I/O Port Assignment Table

The sensors in the information acquisition system are connected to the digital input ports and analog output ports of the PLC, while the actuators are linked to the digital output ports. The integrated water and fertilizer control system features a total of 6 input ports and 7 output ports, all interfaced with the PLC via fieldbus communication. The distribution table of the I/O ports is detailed in Table 7.6.

7.3.3.2 Circuit Design

The electrical system is divided into the main circuit and the control circuit. The main circuit, which operates at 380 V three-phase power, controls the fertilizer pump and other actuators, implementing interlocking through two AC relays. The control circuit, powered by a 220 V AC supply converted to 24 V DC via a switching

Table 7.6 I/O port assignment table

Import		Export	
Name	Interface	Name	Interface
Lower liquid level sensor	I0.0	Irrigation solenoid valve	Q0.0
Upper liquid level sensor	I0.1	A fertilizer solenoid valve	Q0.1
Super water level sensor	I0.2	B fertilizer solenoid valve	Q0.2
PH sensor for mixed fertilizer bucket	AIW0	Acid solenoid valve	Q0.3
EC sensor for mixed fertilizer bucket	AIW2	Fertilizer pump intermediate relay	Q0.4
Soil moisture content sensor	AIW4	Solenoid valve for water replenishment	Q0.5

power supply, powers the PLC, intermediate relays, monitors, sensors, and solenoid valves. Upper-level sensors, lower-level sensors, and water-level sensors provide digital output signals, which are fed into the PLC's digital input ports. In contrast, the soil moisture sensor, pH sensors, and mother liquor tank conductivity sensors generate analog signals. These analog signals are routed to the analog input ports of the Siemens S7-200 SMART EM AE04 module. Within the PLC, these signals are converted to analog engineering registers through internal programs, which subsequently produce control commands for the actuators. To ensure the safety of the PLC controller, intermediate relays are used to connect the PLC with the actuators. This configuration allows for weak power control of strong power components, such as the fertilizer pump, and manages the opening and closing of various solenoid valves, facilitating intelligent irrigation and fertilization. The PLC controller is connected to the HMI interface via Ethernet, providing a real-time, intuitive display of the equipment's operating status and environmental parameters.

7.4 Software Design of Water and Fertilizer Integration System

7.4.1 Software Design Principles and Overall Design Program

7.4.1.1 Overall Software Design Principles

The design of the control system is crucial for ensuring the effective operation of facility-based strawberry water and fertilizer integration equipment. To achieve reliable and stable performance, the following design principles must be considered during software development:

(1) Stability and reliability. The control execution system is fundamental to the equipment's stable operation and safe production. Program design should minimize unnecessary features, enhance program rationality, and reduce interactions between different functional modules to ensure reliability.
(2) Accuracy of data collection. The high temperature and humidity within the greenhouse environment can lead to sensor degradation, affecting data accuracy. To ensure precise system operation, it is essential to use high-precision sensors, increase sensor quantities as needed, and perform regular maintenance and calibration.
(3) Simplicity of the human–computer interface. Given that the primary users are farmers, the interface should be simple and intuitive. A clear, hierarchical, and logically structured interface facilitates ease of use and minimizes the risk of operational errors.
(4) Expandability of the control system. To accommodate future enhancements and additional functionalities, the control system should be designed with expansion capabilities in mind, allowing for future upgrades and modifications.

7.4.1.2 Overall Software Design Program

The overall software design comprises three integral components: the information monitoring and acquisition system, the control and execution system, and the human–computer interaction interface. Together, these components ensure the seamless operation of the water and fertilizer integration system.

Environmental monitoring system software design: This system employs wired connections to continuously monitor and collect equipment operating parameters and greenhouse environmental data. This information is transmitted to the PLC module, where a data conversion program translates the analog signals from the sensors into digital data that the machine can interpret. This digital data provides the necessary input for decision-making within the control system and supports the optimization of the water and fertilizer deployment model by offering critical data insights.

Control and execution system software design: (1) Cooperating with the environmental monitoring system, real-time environmental information and equipment operation information are collected and processed by the internal program of the PLC, outputting digital signals to control the start and stop of solenoid valves, fertilizer pumps, etc., and automatically making decisions on fertilization and irrigation operations. (2) In conjunction with the human–computer interaction interface, users can directly control various execution components of the system through the human–computer interaction interface, and adjust control parameters based on the optimal fertilization and irrigation amount obtained from the previous water and fertilizer allocation model.

Human–computer interaction interface design: (1) In order to more intuitively reflect the operation status of the equipment and the environmental information inside the greenhouse, the human–computer interaction interface should have a system monitoring interface to visualize the data collected by the information collection system, making it easy for users to timely understand the job information; (2) In order to facilitate field management based on actual operational needs, the human computer interaction interface should also have the function of modifying control parameters, timely adjusting parameters such as irrigation amount, fertilizer concentration, and fertilizer type; (3) In order to improve the flexibility of integrated water and fertilizer equipment, the human–machine interface should have the function of switching between manual and automatic operation modes. In automatic mode, the equipment can automatically decide irrigation based on user set parameters. In manual operation mode, users can decide the amount of fertilizer and irrigation and troubleshoot equipment based on the actual growth of crops; (4) In order to ensure the safe operation of the equipment, it is also necessary to set up functions such as abnormal operation status alarm and data recording.

7.4.2 Control System Programming

In this paper, the PLC programming is carried out using STEP 7 MicroWIN SMART software. This software, developed by Siemens for the S7-200 series PLCs, is designed to be compatible with various computer operating systems. It features a user-friendly programming interface that allows for flexible variable name definition and includes a range of functional program blocks. The software supports advanced functionalities such as pulse width modulation and PID control through built-in wizards.

Based on the characteristics of strawberry water and fertilizer demands in southern Xinjiang and the local management practices identified earlier in this study, the control program is designed to encompass several key functionalities: automatic irrigation, automatic fertilization, alternating irrigation and fertilization operations, operational time locking, alarm notifications, and manual automatic mode switching.

7.4.2.1 Analog Conversion Programming

Since the liquid level sensor provides digital information, it can be directly connected to the PLC input port. In contrast, the data from the substrate moisture sensor, fertilizer mixing drum EC sensor, and pH sensor are analog signals. These analog signals need to be converted into engineering units by the PLC system for subsequent processing. The sensor's output current signals in the range of 4–20 mA, which corresponds to a conversion value range of 5300–27,648.

7.4.2.2 Runtime Lock Programming

In practical management, due to the minimal evapotranspiration of plants at night and the absence of on-duty personnel to address unexpected issues, irrigation operations are not conducted during nighttime hours. To address this, the program incorporates a time lock feature. This feature reads the system time, converts it to minutes, and then compares it against a predefined time interval. If the current time exceeds the set irrigation period, the system will refrain from performing fertilization operations.

7.4.2.3 Automatic Irrigation Program Design

In automatic irrigation mode, the fertilizer pump, fertilizer mixing bucket replenishment solenoid valve, and mother liquor bucket solenoid valve remain inactive, with only the irrigation solenoid valve operating. The system sets upper and lower limits for substrate moisture levels. The soil moisture sensor continuously monitors the substrate moisture content. When the substrate moisture falls below the lower limit, the irrigation solenoid valve activates to commence irrigation. Conversely, when the substrate moisture exceeds the upper limit, the irrigation solenoid valve deactivates to halt irrigation. The control flowchart is illustrated in Fig. 7.3.

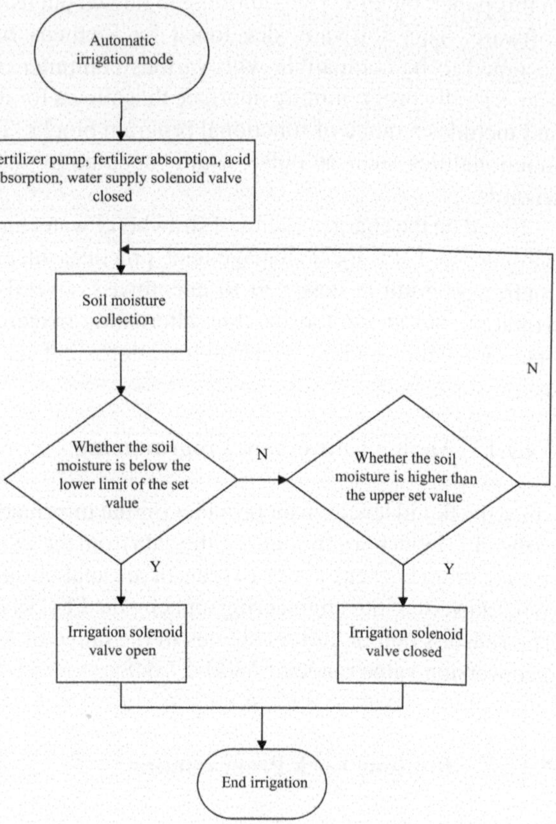

Fig. 7.3 Flowchart of automatic irrigation function

7.4.2.4 Automatic Fertilizer Application Program Design

Strawberry plants exhibit high sensitivity to both water and fertilizer concentrations, making the pipeline mixing method unsuitable due to its poor uniformity and low control accuracy for the specific needs of strawberry facilities. Additionally, substrate conductivity is influenced by water content fluctuations, which can lead to inaccurate reflections of nutrient content and unreliable decisions for fertilizer application. To address these issues, this paper establishes optimal conductivity levels for strawberries during different growth stages as the standard for fertilizer application. The system adjusts the concentration of the mother liquor to the target value in the fertilizer mixing barrel and uses the fertilizer pump to deliver the appropriately mixed solution to the field network, ensuring precise and automated fertilizer application.

In the automatic fertilization operation, the fertilizer barrel must be continuously refilled with water and fertilizer. When the liquid level in the barrel drops below the lower level sensor, the replenishment solenoid valve opens, and the fertilizer pump stops. Once the liquid level reaches the upper-level sensor, this sensor triggers the closure of the replenishment solenoid valve and activates the fertilizer pump for fertilization. If the liquid level is between the upper and lower sensors, the system maintains the replenishment solenoid valve open and the fertilizer pump running, performing fertilization according to preset requirements. Should the liquid level rise to the position of the high water level sensor, the system will trigger an alarm and close all solenoid valves. To account for fluctuations in the liquid level and potential sensor trigger errors due to splashing, a delay is set for the upper, lower, and high water level sensors to prevent actuator malfunctions. Additionally, the fertilizer mixing barrel's EC and pH sensors monitor the solution. If these sensors detect abnormal values, the system triggers an alarm and halts operation. When the conductivity is less than or equal to the set value, corresponding suction solenoid valves are activated based on the selected fertilizer formula, starting the suction operation. The system adjusts the concentration of the mixed liquid: if the actual concentration exceeds the set value, the suction solenoid valve closes. To regulate pH, if the value is higher than the set point, the acid-absorbing solenoid valve opens; if lower, the valve closes. The flowchart for fertilizer barrel replenishment is illustrated in Fig. 7.4.

7.4.2.5 Fertilizer Application Program Design for Recirculating Irrigation

Strawberry growth is suitable for thin fertilizer hard application, this equipment is set up for fertilizer, water cycle times proportion, when the water set number of times the end, switch to fertilizer irrigation, and so on and so forth.

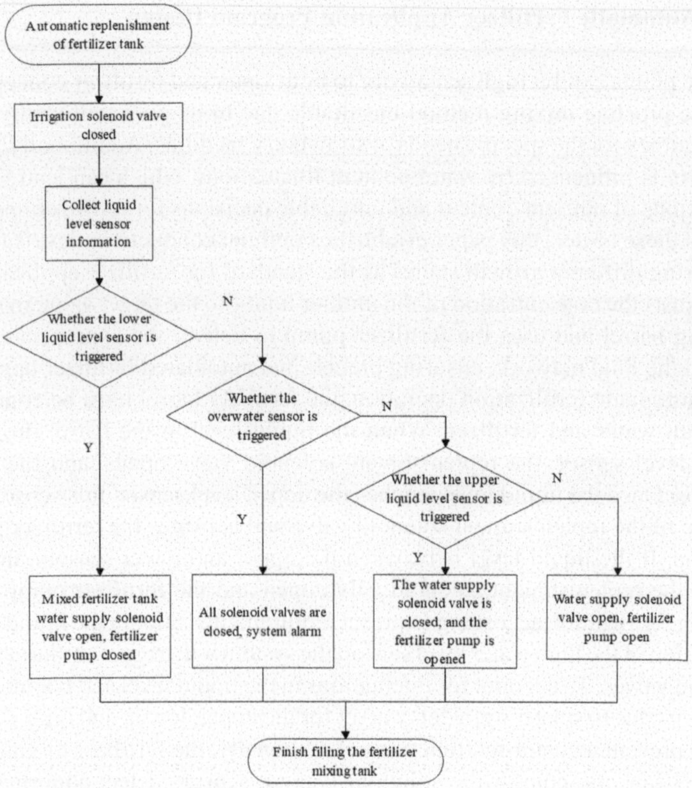

Fig. 7.4 Fertilizer drum replenishment function flow chart

7.5 Human–Computer Interface Software Design

HMI is an intuitive interface that connects people and equipment. Users can send control commands to the lower unit through the HMI page, and the lower unit can also visualize the equipment operation to the user, and the HMI page selects Haiwell C7S-GW.

The display combines ordinary HMI with IOT gateway and DTU module, which drastically reduces the installation space and hardware cost, and is characterized by simple programming, low failure rate and long service life. At the same time, this HMI HMI also integrates Haiwell Cloud service and built-in Haiwell cloud engine, which can directly map the HMI interface to the mobile terminal, and drastically reduce the Remote control interface programming time is greatly reduced. The HMI is connected to the PLC via Ethernet, the system operation interface is written by Haiwell Cloud Scada Designer configuration software.

The human–computer interaction interface comprises several key components: the startup interface, automatic setting interface, manual setting interface, intelligent irrigation interface, and historical data query interface. Upon powering on the water and fertilizer integration equipment, users are greeted by the startup interface. In the startup interface, users can monitor the equipment's operational status in real time. If a system failure occurs, the equipment halts, all solenoid valves close, the fertilizer pump ceases operation, and an alarm buzzer sounds. Users can reset the system and resume normal operations by pressing the alarm reset button after troubleshooting the issue. Additionally, the interface features a pause button for users to temporarily stop the equipment as needed. The startup interface includes a row of three fertilizer selection buttons, allowing users to choose the appropriate fertilizer type based on their specific requirements. Moreover, the interface enables users to set desired values for fertilizer conductivity and pH. It displays the real-time conductivity and pH values within the fertilizer barrel, providing an intuitive overview of the current fertilizer conditions.

In order to avoid sensor acquisition data errors, in the solar greenhouse uniformly set up three soil sensors, real-time collection of soil moisture, temperature and conductivity, and take the average value of the three sensors as the input to the system, in the soil moisture stabilization zone were set at the upper limit of the soil moisture and the lower limit, when the soil moisture content is lower than the lower limit of moisture, the system automatically carry out the irrigation operation When the soil moisture content is lower than the lower limit, the system will automatically irrigate, and when the soil moisture content is higher than the upper limit, the irrigation will stop; soil conductivity and humidity are mainly used for the user to understand the situation in the greenhouse more intuitively, and do not participate in the system decision-making. At the same time, the user can also according to the actual planting needs, by changing the ratio of water and fertilizer, to determine the number of cycles of water and fertilizer application; due to the high humidity at night, and unmanned, equipment problems can not be solved, so set the equipment running time period, to avoid the night operation.

The historical data query interface, allows users to retrieve information on the number of equipment operations, the conductivity of the output fertilizer solution, and other relevant parameters within a specified time period. By inputting the start and end times, users can gain insights into the fertilizer application and irrigation activities during that period.

7.6 Research on Fuzzy PID Control Strategy

Fuzzy PID control strategies are employed in water-fertilization systems to address the complexity and uncertainty inherent in crop growth environments. While conventional PID control is commonly used in industry, it often suffers from limitations such as insufficient control accuracy and slow response speed, especially when managing the nonlinear variations characteristic of agricultural settings. The fuzzy PID control strategy offers a more adaptive solution by integrating the benefits of fuzzy logic with traditional control methods, enhancing both accuracy and responsiveness in dynamic and unpredictable environments.

The application of the fuzzy PID control strategy in this system is notably demonstrated in the real-time adjustment of irrigation and fertilization parameters. The water and fertilizer integration system continuously monitors environmental parameters such as soil moisture, conductivity, and temperature through sensors. These parameters fluctuate in response to changes in crop growth stages and external climatic conditions. The fuzzy PID control strategy dynamically adjusts the PID controller parameters—proportional coefficient (Kp), integral coefficient (Ki), and differential coefficient (Kd)—based on real-time data to ensure precise irrigation and fertilization. For instance, if the system detects that soil moisture falls below the set threshold, the fuzzy controller can quickly increase the irrigation volume. Conversely, if conductivity deviates from the predetermined range, the controller can promptly adjust the fertilizer concentration to prevent over- or under-fertilization. This strategy also significantly enhances the response speed and stability of the system. By adjusting control parameters in real-time, the system maintains optimal performance across varying conditions, which is essential for effective water and fertilizer management. In the dynamic agricultural environment, the adaptive nature of the fuzzy PID control strategy minimizes overshooting, improves system stability, and ensures a continuous and stable supply of water and fertilizer to the crops.

7.6.1 Principles of Fuzzy PID Control

The conventional PID control formula, as shown in Eq. (7.1), calculates the control output by processing the proportional, integral, and differential components of the error with fixed weighting factors. However, this method relies on static PID parameters from the previous cycle and cannot adapt to changes in the actual working environment. As a result, it may suffer from reduced control accuracy and stability. To address these limitations, it is necessary to introduce correction factors to dynamically adjust the three PID parameters (proportional, integral, and differential), thereby improving the control system's adaptability and performance.

$$u(t) = kp_0 e(t) + ki_0 \int_0^t ue(t)dt + kd_0 \frac{de(t)}{dt} \tag{7.1}$$

Fuzzy control has the principle of fuzzy control is to take the error e of the EC value of water and fertilizer conductivity and the rate of change ec of the error e as the input of the system, and the input variables are converted from exact variables to discrete variables after fuzzy processing, and the fuzzy control quantities are outputted after the fuzzy rule reasoning written by artificial experience. Through the defuzzification process, the final output of three tuning parameters Δkp, Δki, Δkd, each tuning parameter is respectively multiplied by the corresponding proportionality factor, and summed up with the PID parameters before tuning to obtain the rectified

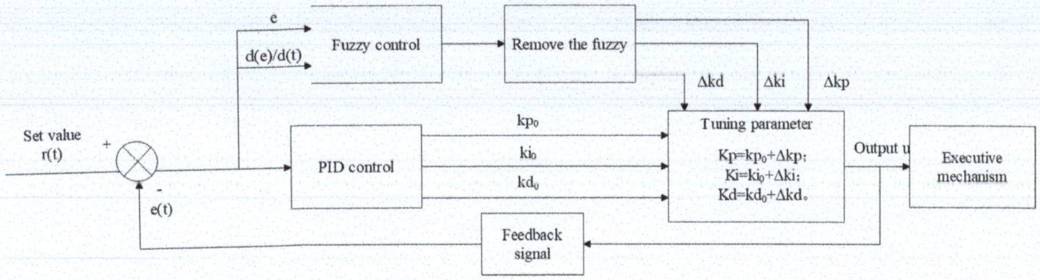

Fig. 7.5 Fuzzy PID control principle

system parameters, i.e., the fuzzy PID control parameters, which realizes the self-tuning of the PID parameters as shown in Eq. (7.2), and ultimately realizes the closed-loop control regulation of the control system. Fuzzy PID control strategy is essentially a combination of fuzzy control and PID control, the control process is shown in Fig. 7.5.

$$\begin{cases} Kp = \Delta kp \times x_p + kp_0 \\ Ki = \Delta ki \times x_i + ki_0 \\ Kd = \Delta kd \times x_d + kd_0 \end{cases} \tag{7.2}$$

In Eq. (7.2), Kp represents the proportional coefficient after tuning; ki represents the integral coefficient after tuning; kd represents the differential coefficient after tuning; Δkp represents the proportional correction factor; Δki represents the integral correction factor; Δkd represents the differential correction factor; x_p represents the proportional factor for the proportional coefficient; x_i represents the proportional factor for the integral coefficient; x_d represents the proportional factor for the differential coefficient; kp_0 represents the proportional coefficient before tuning; ki_0 represents the integral coefficient before tuning; kd_0 represents the differential coefficient before tuning.

7.6.2 *Design of the Fuzzy Controller*

The design of the fuzzy controller is mainly divided into two parts, the first part is to set the fuzzy linguistic values and fuzzy domain of the input and output variables of the system; the second part is to write the expert experience into fuzzy rules to form a fuzzy set.

Firstly, the deviation e of the EC value of water and fertilizer conductivity and the rate of change of the deviation of e, ec, are taken as the two input parameters of the 2D fuzzy control system, and since the fuzzy rules cannot identify the exact variables, they need to be discretized, and the discretized fuzzy domains of e and ec are taken as:

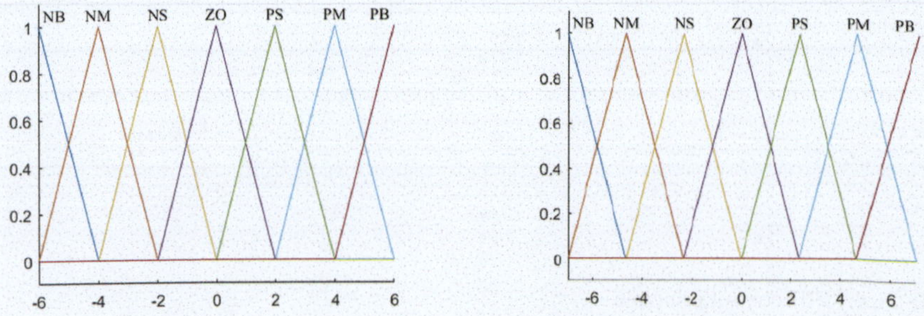

a: *e*, *ec* affiliation function curves

b: curve of the Δkp, Δki, Δkd subordination function

Fig. 7.6 Affinity function curve

$$e = ec = \{-6, -5, -4, -3, -2, -1, 0, 1, 3, 4, 5, 6\} \tag{7.3}$$

The linguistic values of *e* and *ec* are {NB (Negative Big), NM (Negative Medium), NS (Negative Small), ZO (Zero), PS (Positive Small), PM (Positive Medium), PB (Positive Big)}. The fuzzy subsets of *e* and *ec* use the triangular affiliation function, and the affiliation curves are shown in Fig. 7.6a.

The three output parameters Δkp, Δki, Δkd of the two-dimensional fuzzy control system take the values of the fuzzy theory domain after discretization:

$$\Delta kp = \Delta ki = \Delta kd = \{-7, -6, -5, -4, -3, -2, -1, 0, 1, 3, 4, 5, 6, 7\} \tag{7.4}$$

The linguistic values of Δkp, Δki, and Δkd are {NB (negative large), NM (negative medium), NS (negative small), ZO (zero), PS (positive small), PM (positive medium), PB (positive large)}.

Δkp, Δki, and Δkd three output parameter fuzzy subsets are selected as triangular affiliation functions, and the affiliation curves are shown in Fig. 7.6b.

According to the magnitude of the system input deviation *e* and deviation change rate *ec*, the corresponding control quantities are output, according to which 49 fuzzy control rules are written:

(1) If (*e* is NB) and (*ec* is NB) then (*kp* is PB) (*ki* is NB) (*kd* is PS);

(2) If (e is NB)and(ec is NM)then(kp is PB) (ki is NB) (*kd* is PS);

...

(49) If (*e* is PB) and (*ec* is PB) then (*kp* is NB) (*ki* is PB) (*kd* is PB).

The fuzzy control rules are shown in Table 7.7, where the language values are Δkp, Δki, and Δkd from left to right.

Δkp, Δki, Δkd fuzzy rules three-dimensional surface is shown in Fig. 7.7, the rule surface from the highest point to the lowest point of the transition is smooth, there is no mutation rule that does not meet the overall trend of change, so it can be seen that the written control rules can achieve a better control effect.

Table 7.7 Fuzzy rule control table

Project		Error *e* linguistic value						
		NB	NM	NS	ZO	PS	PM	PB
Error rate of change *ec* langu-age value	NB	PB/NB/PS	PB/NB/NS	PM/NM/NB	PM/NM/NB	PS/NS/NB	ZO/ZO/NM	ZO/ZO/PS
	NM	PB/NB/PS	PB/NB/NS	PM/NM/NB	PS/NS/NM	PS/NS/NM	ZO/ZO/NS	NS/ZO/ZO
	NS	PM/NB/ZO	PM/NM/NS	PM/NS/NM	PS/NS/NM	ZO/ZO/NS	NS/PS/NS	NS/PS/ZO
	ZO	PM/NM/ZO	PM/NM/NS	PS/NS/NS	ZO/ZO/NS	NS/PS/NS	NM/PM/NS	NM/PM/ZO
	PS	PS/NM/ZO	PS/NS/ZO	ZO/ZO/ZO	NS/PS/ZO	NS/PS/ZO	NM/PM/ZO	NM/PB/ZO
	PM	PS/ZO/PB	PS/ZO/NS	NS/PS/PS	NM/PS/PS	NM/PM/PS	NM/PB/PS	NB/PB/PB
	PB	ZO/ZO/PB	ZO/ZO/PM	NM/PS/PM	NM/PM/PM	NM/PM/PS	NB/PB/PS	NB/PB/PB

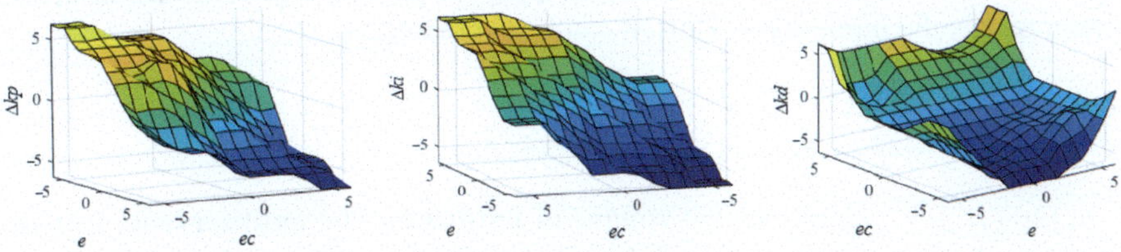

a: Δkp fuzzy regular 3D surface b: Δki fuzzy regular three-dimensional surface c: Δkd fuzzy regular 3D surface

Fig. 7.7 Fuzzy rule 3D surface map

According to the fuzzy rule control table, the weighted average method is used to defuzzify, and the expression of the area center of gravity method is shown in Eq. (7.5), and finally the defuzzified values of Δkp, Δki, and Δkd are obtained.

$$\Delta kp = \frac{\sum_{j=1}^{49} u_p(\Delta k_{pj}) \Delta k_p}{\sum_{j}^{49} u_p(k_{pj})} \tag{7.5}$$

In Eq. (7.5), $u_p(\Delta k_{pj})$ is the affiliation function of the fuzzy set of output quantities.

By calculating the error e and the error change rate ec between the target value and the actual value, combined with fuzzification, fuzzy reasoning and defuzzification, the final loss PID parameter increments Δkp, Δki, Δkd, the parameter increments and the initial value of the PID can be added together to get the control quantity of the next moment as shown in Eq. (7.6).

$$u(t) = (\Delta kp + kp_0)e(t) + (\Delta ki + ki_0)\int_0^t e(t)dt + (\Delta kd + kd_0)\frac{de(t)}{dt} \tag{7.6}$$

7.6.3 *Simulation Analysis*

The control strategy gathers data on the conductivity of the fertilizer solution in the mixing drum via sensors, and then issues control commands based on a comparison with preset values to regulate the amount of fertilizer intake, thereby achieving dynamic adjustment of the fertilizer solution's concentration. However, in the actual adjustment process, numerous uncertain interfering factors are present, making it difficult to represent these factors through an accurate mathematical model. Therefore, this paper analyzes the variation in water and fertilizer concentration by establishing a simplified fertilizer mixing model.

When the system reaches equilibrium, the volume in the fertilizer mixing tank remains constant. At this point, the system adheres to the principles of mass conservation and flow conservation, meaning that the output mass of fertilizer from the system equals the input mass. Thus, it can be expressed as:

$$\frac{d[V_T C(t)]}{dt} = C_1 q_1(t) + C_2 q_2(t) - C(t) q_3(t) \tag{7.7}$$

In Eq. (7.7), V_T represents the volume of fertilizer in the fertilizer tank, L; $C(t)$ represents the concentration of output fertilizer, g/L; C_1 represents the concentration of input mother liquor, g/L; q_1 represents the flow rate of input mother liquor, L/s; C_2 represents the concentration of input water, g/L; q_2 represents the flow rate of input water, L/s; q_3 represents the flow rate of output fertilizer, L/s.

Also, since conductivity is proportional to the concentration of water and fertilizer, it can be obtained by replacing the liquid concentration with the value of conductivity Ec:

$$\frac{d[V_T Ec(t)]}{dt} = Ec_1 q_1(t) + Ec_2 q_2(t) - Ec(t) q_3(t) \tag{7.8}$$

In Eq. (7.8), $Ec(t)$ represents the output mixed fertilizer liquid conductivity value, mS/cm; Ec_1 represents the input mother liquor conductivity value, mS/cm; Ec_2 represents the input clear water conductivity value, mS/cm.

Equation (7.8) are obtained by performing a Rasch transformation:

$$E(s) = \frac{Ec_1 q_1 + Ec_2 q_2}{V_T s + q_3} \tag{7.9}$$

In Eq. (7.9), s represents the complex variable of the Raschel transform. From Eq. (7.9), the control response is characterized as a first-order linear system, and let the whole system Ec_1 be 15 mS/cm, Ec_2 be 1.43 mS/cm, V_T be 70 L, q_1 be 0.055 L/s, q_2 be 2 L/s, and q_3 be 2 L/s. Due to the existence of a certain hysteresis during the actual control process, let the hysteresis time be 7 s. Then the approximate transfer function of the system is:

$$G(s) = \frac{3.685}{70s + 2} e^{-7s} = \frac{1.8425}{35s + 1} e^{-7s} \tag{7.10}$$

Simulink module in Matlab R2022b version is selected as the simulation software, and the simulation hardware CPU is Intel Core i5-7300HQ, and the simulation model is shown in Fig. 7.8, which is set up to simulate the traditional PID control strategy and the fuzzy PID control strategy, respectively. The proportional, integral and differential coefficients of the control strategy are derived using the critical ratio method, and then the quantization factor and the proportionality factor of the error e and the error change rate ec are obtained according to the empirical trial-and-error method.

Respectively, traditional PID control and fuzzy PID control will be used to conductivity to 1.93, 2.13, 2.34 mS/cm; simulation sampling time is set to 300 s, and the rest of the parameters remain unchanged, the simulation results are shown in Table 7.8.

The simulation results show that the average steady state error of the conventional PID control strategy to regulate the target conductivity to the three set values is 0.018 mS/cm, the average steady state time is 318 s, and the average overshoot is 12.87%. The fuzzy PID control strategy regulates the target conductivity to three target values

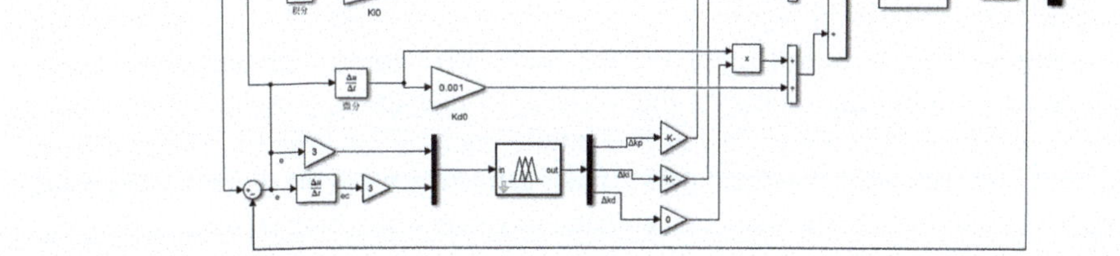

Fig. 7.8 Simulink simulation model of fuzzy PID control strategy

Table 7.8 Simulation comparison of two control strategies

Control mode	EC set value (mS cm^{-1})	Steady-state error (mS cm^{-1})	Resting time (s)	Overshoot (%)
Traditional PID control	1.93	±0.015	265	12.13
	2.13	±0.020	318	12.85
	2.34	±0.020	372	13.63
Fuzzy PID control	1.93	±0.010	188	3.17
	2.13	±0.010	236	5.22
	2.34	±0.010	262	6.39

with an average steady-state error of 0.01 mS/cm, an average steady-state time of 229 s, and an average overshoot of 4.92%.

As the set conductivity increases, the time required to reach a steady state for both control strategies also increases. However, the change in steady-state error remains relatively small. At the same time, the fuzzy PID control strategy exhibits a smaller steady-state error and less overshoot compared to the traditional PID control strategy. This indicates that the fuzzy PID control strategy offers superior control performance over the traditional PID approach.

7.7 Conclusions of the Study on Smart Water and Fertilizer Integration Systems for Desert Gobi Facilities

In this study, we proposed and developed an intelligent water and fertilizer integration system to address the challenges of water scarcity, poor soil conditions, and the inefficiencies of traditional agriculture faced by facility-based farming in the desert Gobi region. To tackle these issues, we constructed a water and fertilizer deployment model tailored for facility agriculture in the desert Gobi, taking into account the unique environmental conditions of Southern Xinjiang. The model fully considered key environmental parameters such as soil moisture, temperature, and conductivity, and determined water and fertilizer supply strategies for different crop growth stages, providing a reference for precise system control. By incorporating a fuzzy PID control strategy, the intelligent water and fertilizer integration system developed in this study is capable of real-time environmental monitoring and dynamic adjustment of irrigation and fertilization parameters based on the actual needs of crops. Compared to traditional PID control methods, the fuzzy PID control strategy demonstrates better adaptability and control effectiveness in managing complex environmental changes, thus achieving precision and intelligence in water and fertilizer management. Field tests confirmed that the system performs well in the desert Gobi environment, enhancing the efficiency of water and fertilizer utilization while reducing water waste. Additionally, the system positively impacts crop yield and quality. The reduction in manual operations through automated control further improves the efficiency of agricultural production and management.

The intelligent water and fertilizer integration system developed in this study offers a valuable solution for facility agriculture in the desert Gobi region, enhancing the effectiveness of water and fertilizer management. This system not only provides a reference for similar agricultural environments but also offers technical support for future agricultural production in comparable conditions.

Chapter 8
Study on Spray Agent Model for Gobi Facilities in Desert

Long Wang, Xufeng Wang, Can Hu, Yi Zhu, and Guansan Zhu

8.1 Introduction

Xinjiang, located in the northwest of China, is divided into southern and northern regions by the Tianshan Mountains. Southern Xinjiang is centered around the Taklamakan Desert, surrounded by mountains on three sides, characterized by an arid, low-precipitation temperate climate. However, its abundant soil, water, and sunlight resources are favorable for agricultural and pastoral development. At present, greenhouse-grown tomatoes have become an important economic crop in southern Xinjiang, making the scientific and effective management of modern greenhouse facilities essential. One crucial element in ensuring high yields and quality of tomatoes is plant protection operations.

To improve greenhouse spraying technology, this research integrates advanced electrostatic spraying techniques into existing plant protection machinery. A prototype electrostatic sprayer specifically designed for greenhouse tomatoes has been developed, with core components designed accordingly. Field measurements conducted in Alar City, First Division, show that the height of mature tomato plants in greenhouses is around 180 cm, with varying plant densities in different canopy layers. The row spacing between crops is approximately 70 cm, and the base height of the plants is about 60 cm. Therefore, the height of the sprayer boom was set at 120 cm. There has been limited research on the performance of vertical electrostatic

L. Wang · X. Wang (✉) · C. Hu · Y. Zhu · G. Zhu
College of Mechanical and Electrical Engineering, Tarim University, Alar 843300, China
e-mail: wxfwyq@126.com

Modern Agricultural Engineering Key Laboratory at Universities of Education Department of Xinjiang Uygur Autonomous Region, Alar 843300, China

Xinjiang Production and Construction Corps (XPCC) Key Laboratory of Utilization, Equipment of Special Agricultural and Forestry Products in Southern Xinjiang, Alar 843300, China

L. Wang
e-mail: 120140002@taru.edu.cn

C. Hu et al. (eds.), *New Technologies Applied in Gobi Desert Facility Agriculture*, Smart Agriculture 11, https://doi.org/10.1007/978-981-96-1031-0_8

sprayers in current literature. In this chapter, the distribution of the electrostatic field in the surrounding area of electrostatic nozzles is simulated using finite element analysis software, Ansoft Maxwell. The simulation is based on parameters like nozzle spacing and electric ring settings, providing insights into the electrostatic field characteristics, which will serve as the foundation for future field trials of vertical boom electrostatic sprayers.

8.2 Analysis of Tomato Growth Environment

8.2.1 Greenhouse Structure

The greenhouse vegetable cultivation patterns vary across different regions, and the design of agricultural machinery must be integrated with agronomic practices. Therefore, to develop a greenhouse sprayer suitable for the planting systems in southern Xinjiang, it is necessary to conduct a thorough investigation of the local greenhouse layouts. Based on the tomato cultivation methods in southern Xinjiang's facilities, an analysis of the greenhouse environment should be conducted, along with a detailed investigation of crop characteristics such as row width, number of rows, and growth patterns. This will guide the design of the equipment's operational parameters to ensure effective crop protection for greenhouse tomatoes. The cultivation pattern of tomatoes in greenhouses differs from that in standard greenhouses, requiring the construction of wire trellises. Figure 8.1 shows the structure of a tomato trellis greenhouse in the 1st Division of Aral, which includes components such as inter-row operational paths, main walkways, support pillars, hanging wires, and wire trellises. Relevant dimensions are listed in Table 8.1.

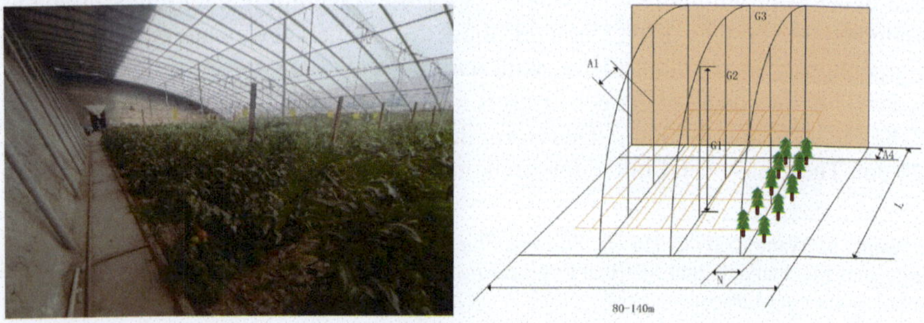

Fig. 8.1 The greenhouse structure of tomato vines in Alaer City, the First Division

Table 8.1 Greenhouse support structure parameters

	Horizontal distance (m)				Vertical distance (m)		
Size code	N	A1	L	A4	G1	G2	G3
Specific parametes	0.7	4	14	0.8	1.5	3	8

8.2.2 *Growth Characteristics of Indoor Tomatoes*

Tomato is a perennial herb, and China ranks among the world's top producers of tomatoes while also being one of the largest consumers. Aral City, located between the Kunlun Mountains to the north and the Tianshan Mountains to the south, benefits from an exceptional geographical environment, which provides favorable conditions for the development of facility agriculture. In recent years, with the rapid growth of facility agriculture, the number of solar greenhouses in Aral City has steadily increased. As one of the main fruit-bearing vegetable crops, tomatoes hold a significant position in the overall vegetable industry. The construction of greenhouses in Aral's farming communities employs a brick wall and steel beam structure. These greenhouses typically span 12–13 m from north to south and 140–160 m from east to west. They are covered with PVC plastic films, known for their excellent light transmission, strong insulation properties, and long service life. According to preliminary research, when tomatoes are first planted, their height ranges from 20 to 30 cm. A double-row planting method is generally used, with approximately 2,000 plants per mu (a Chinese unit of area). The row spacing for tomato plants is about 70 cm, with an intra-row spacing of about 35 cm. Measurements show that tomato plants are generally maintained at a height of around 180 cm. Based on the plant height, the working space within the greenhouse can be divided into upper, middle, and lower sections.

8.3 Electric Field Simulation Study

8.3.1 *Introduction to Maxwell Electric Field Simulation Software*

Ansoft Maxwell, developed by Ansoft Corporation, is a software tool used for both 2D (Maxwell 2D) and 3D (Maxwell 3D) electromagnetic field analysis. Based on Maxwell's equations, the software offers several advantages compared to other finite element analysis tools: (1) User-friendly interface: the intuitive graphical interface allows for easy model creation directly within the platform. (2) Fast computation: Maxwell has a high-speed processing capability and comprehensive data handling features. (3) Flexible meshing: Users can customize the mesh or perform local optimization to suit their analysis needs. (4) Versatile analysis: The software supports a wide range of linear and nonlinear analyses. As the demands for improved product performance and more compact designs become increasingly stringent, the requirements for digital field simulation technology have also risen. This has allowed Maxwell to stand out from its competitors, thanks to its practicality and powerful digital functionalities.

8.3.2 *Finite Element Principles of Ansoft Maxwell*

In the mid-19th century, British physicist James Clerk Maxwell developed a mathematical model to describe macroscopic electromagnetic phenomena—known as Maxwell's equations. These equations consist of four fundamental laws: Ampère's Circuital Law: The line integral of the magnetic field intensity around any closed loop equals the total current passing through the surface Ω defined by the loop. The mathematical expression is:

$$\oint_r H \cdot dl = \iint_S J \cdot dS + \iint_\Omega \frac{\delta D}{\delta t} \cdot dS \tag{8.1}$$

Its differential form is:

$$\nabla \cdot \vec{H} = \vec{J} + \frac{\delta \vec{D}}{\delta t} \tag{8.2}$$

where, J is the current density vector, $\frac{\delta D}{\delta t}$ is the displacement current density, A/m^2; D is the electric displacement vector, C/m^2; ∇ represents the vector differential operator.

Gauss's Law for Electricity: The electric flux through any closed surface is proportional to the charge enclosed by that surface. The mathematical expression is:

$$\oiint_A D \cdot d\vec{s} = \oiint_v \rho \, dv \tag{8.3}$$

Its differential form is:

$$\nabla \cdot \vec{D} = \rho \tag{8.4}$$

where, V is the volume enclosed by the surface; ρ is the charge density. Gauss's Law for Magnetism: In a magnetic field, regardless of the distribution of the medium or the magnetic flux density, the total magnetic flux through any closed surface is always zero. The mathematical expression is:

$$\oiint_S B \cdot dS = 0 \tag{8.5}$$

Its differential form is:

$$\nabla \cdot B = 0 \tag{8.6}$$

where, B represents the magnetic flux density, T.

Faraday's Law of Electromagnetic Induction: The electromotive force (EMF) induced in a closed loop is proportional to the rate of change of the magnetic flux through the loop. The mathematical expression is:

$$\int_r E \cdot dl = -\iint_\Omega \frac{\delta \vec{B}}{\delta t} \cdot d\vec{S} \tag{8.7}$$

In this simulation, the 3D electric field solver used follows the equation:

$$\Delta \cdot (\varepsilon_r \cdot \varepsilon_0 \nabla \Phi) = -\rho_V \tag{8.8}$$

where, $\Phi(x, y, z)$ is the three-dimensional scalar electric potential, $\varepsilon_r(x, y, z)$ is the relative permittivity along the three vector directions, ε_0 is the permittivity of free space, $\rho_V(x, y, z)$ is the volume charge density.

The 3D scalar potential Φ needs to be solved in Maxwell 3D's electric field module. From Maxwell's differential equations, the electric field intensity E and the electric displacement vector D, are then derived using the following fundamental equations.

$$\begin{aligned} E &= -\nabla\Phi \\ D &= \varepsilon_r \cdot \varepsilon_0 \cdot (-\nabla\Phi) \end{aligned} \tag{8.9}$$

8.3.3 *Ansoft Maxwell Calculation Process*

The Ansoft Maxwell software follows a relatively standardized workflow for electric field simulation analysis. The main steps of the process are as follows:

(1) Select the solver type.
(2) Build the simulation model.
(3) Set material properties.
(4) Set excitation sources and boundary conditions.
(5) Set adaptive mesh refinement.

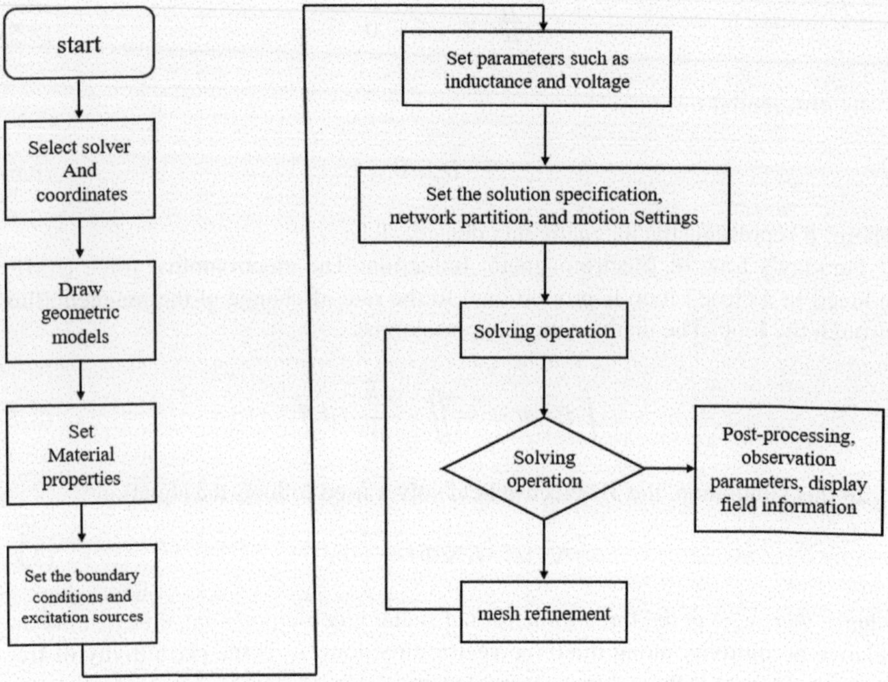

Fig. 8.2 Electric field simulation calculation flow chart

(6) Perform finite element simulation.
(7) Post-process the simulation results.

A flowchart illustrating this simulation process is shown in Fig. 8.2.

8.4 Electrostatic Spray Simulation Model and Factors Affecting Field Strength

8.4.1 Simulation Model Establishment

The finite element analysis software Ansoft Maxwell is used to simulate the electric field around an electrostatic nozzle. In this study, the design employs inductive charging, where the electrostatic field is formed by an annular electrode and a nozzle. Following the Ansoft Maxwell simulation workflow outlined in Sect. 8.3.3, the modeling process is carried out, and the physical model for a single nozzle is shown in Fig. 8.3. The materials for both the nozzle and the electrode are set to steel-1008. The nozzle is simplified as a cylinder with a radius of r, cylinder with a length of 18 mm, The electrode has a diameter of D, a thickness of h, and a width L2 =

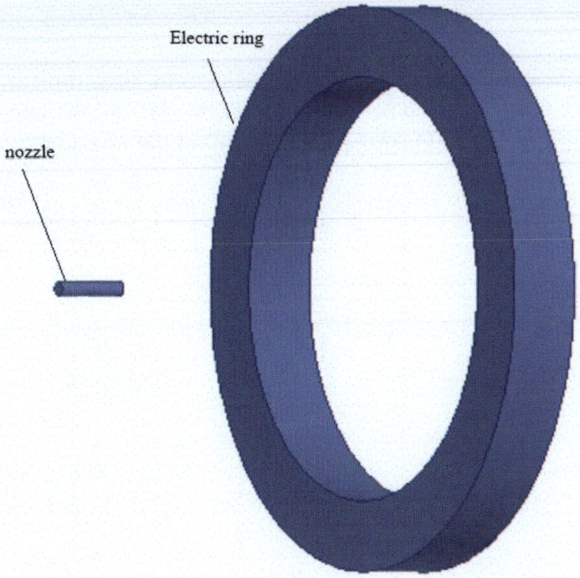

Fig. 8.3 Physical model of a single sprinkler

10 mm, The distance between the nozzle and the electrode is L, with the coordinate origin located at the center of the electrode, The solver type chosen is the Electrostatic Solver, and the solution domain is set to air. The calculation region is selected based on the operational range of the vertical spray boom, with dimensions of 500 mm × 500 mm × 2000 mm. This is aligned with the operational boundary conditions of the sprayer to ensure the simulation's accuracy and reliability, facilitating observation of the electric field distribution.

8.4.2 *Factors Affecting Electrostatic Field Strength*

In the vertical spray rod electrostatic spraying system, the surface charge density ρ_s (in C/m^2) of the liquid can be calculated using Gauss's Law:

$$\rho_s = \varepsilon E \tag{8.10}$$

where, ε is the permittivity of air, F/m; E is the electric field strength, V/m.

The spray current I (μA) is calculated by:

$$I = 2\pi r \rho s v = 2\rho s q V / r = 2\varepsilon E q V / r \tag{8.11}$$

where, r is the nozzle radius, mm; v is the exit velocity, m/s; q_V is the spray volumetric flow rate, m^3/s.

The charge on the annular electrode is given by:

$$Q = 4\pi\varepsilon DU \tag{8.12}$$

where: D is the diameter of the electrode ring, mm; U is the applied voltage, kV.

If the distance between the nozzle exit and the annular electrode is h (mm), then the electric field strength E_∞ (V/m) at the nozzle exit can be expressed as:

$$E = \frac{1}{4\pi\varepsilon}\frac{Qh}{\left(h^2 + \frac{D^2}{4}\right)^{3/2}} = \frac{DUh}{\left(h^2 + \frac{D^2}{4}\right)^{3/2}} \tag{8.13}$$

Substituting Eq. 8.13 into Eq. 8.11 gives:

$$I = \frac{2\varepsilon q_v DUh}{r\left(h^2 + \frac{D^2}{4}\right)^{3/2}} \tag{8.14}$$

The charge-to-mass ratio (μC/kg) of the spray droplets is given by:

$$R_{cm} = \frac{1}{\rho q_v} = \frac{2\pi DUh}{\rho r\left(h^2 + \frac{D^2}{4}\right)^{3/2}} \tag{8.15}$$

where, ρ is the density of water, kg/m^3.

From Eq. 8.15, it is evident that for a given nozzle, the charge-to-mass ratio depends on the applied voltage U, the diameter of the electrode ring D, and the electrode distance h, The charge on the droplets primarily occurs in the area around the nozzle electrode. To achieve optimal charging efficiency and improve the performance of the electrostatic spraying, it is crucial to analyze the impact of parameters such as the applied voltage U and the electrode distance h on the electric field distribution when designing the vertical spray rod electrostatic device.

8.5 Calculation Results and Analysis

8.5.1 *Vertical Multi-Nozzle Electrostatic Field Distribution Law*

Based on the previously described simulation model of a single electrostatic nozzle, we establish the electrostatic field model for a vertical spray rod nozzle and annular electrode. According to Gauss's Law, the maximum electric field strength is expected to occur at the end of the jet. This is because the charge distribution is most concentrated at the jet end. As droplets pass through this region, they experience the maximum electric field force and, thus, are charged most effectively. To analyze the electric field distribution, we select a vertical plane view for observation. In this subsection, the XOZ plane is chosen with the coordinate origin at the tip of the central nozzle. Line segments are drawn on this plane, with starting coordinates at (0, 0, 420) and (0, 0, −420). The electrostatic field simulation results along this line segment are shown in Fig. 8.4. From the figure, it can be observed that the electric field strength around the vertical electrostatic nozzle is symmetrically distributed. Near the nozzle tip, the electric field strength is higher, and it increases gradually as the distance from the nozzle center decreases, appearing as red regions. This phenomenon is typically due to the relationship between the electric field strength and distance, as well as the visualization techniques used. In the XOZ plane, the field strength distribution along the z-axis is examined with z ranging from −420 to 420 mm. As shown in Fig. 8.5, the electric field strength values along this line segment for various nozzles exhibit a trend of initially stabilizing, then suddenly increasing, and finally decreasing. The maximum electric field strength values are observed at positions m1, m2, and m3. From these observations, it is evident that within the entire atomization zone, the fragmentation and charging of the liquid primarily occur near the nozzle tip. Additionally, the electric field strength E on either side of the nozzle decreases as |z| increases.

In the YOZ plane, the electric field distribution varies on either side of the annular electrode. As shown in Fig. 8.6, the electric field strength E exhibits an asymmetric distribution along the y-axis. At 5 mm, which serves as the boundary point between the left and right sides of the electrode ring, the electric field strength on the left side is higher than on the right side. From the nozzle to the annular electrode, the electric field strength along the axis rapidly decreases, reaching its minimum at the center of

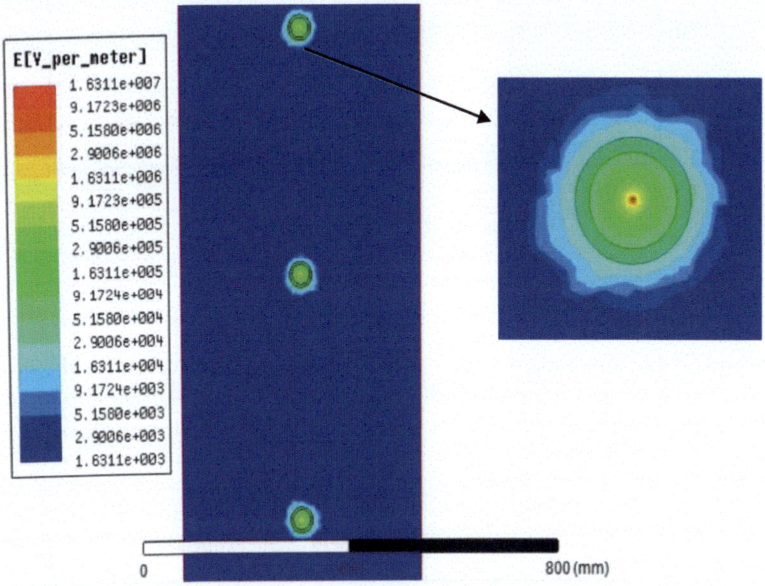

Fig. 8.4 Vertical spray boom electrostatic simulation diagram

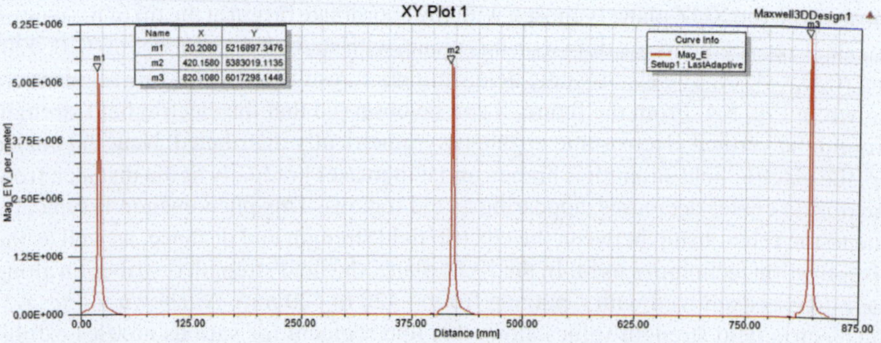

Fig. 8.5 Radial distribution of electric field intensity of multiple nozzles

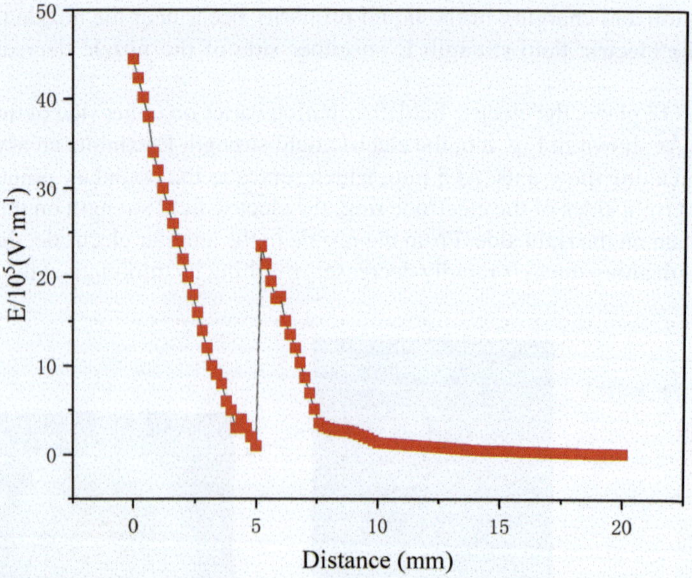

Fig. 8.6 Axial distribution of electric field strength of multiple nozzles

the electrode ring. This indicates that the effective charging region for the droplets within the electric field is quite small. On the right side of the annular electrode, the electric field strength initially increases to a certain value and then gradually decreases.

8.5.2 *Effect of Electrode Voltage on Electric Field Distribution of Electrostatic Sprinkler*

During the operation of an electrostatic spray nozzle, it is surrounded by air, which typically does not conduct electricity under normal conditions. However, at high voltages, the free electrons in the air gain sufficient energy and become ionized, allowing the air to conduct electricity. Therefore, the charging voltage of the electrostatic spray electrodes significantly affects the stability of the electric field distribution. Figure 8.7 shows the voltage-current characteristic curve for gas conductivity. The curve indicates that as the charging voltage increases, the current initially rises, then stabilizes, and eventually increases sharply. This suggests that higher charging voltages are not always better. When the voltage becomes too high, the air medium is ionized, and the charging mode shifts to corona discharge, which reduces charging stability. Therefore, a voltage range of 0 to U2 is more suitable, with U1 being the optimal value. Since the electrostatic spray utilizes inductive droplet charging, it is crucial to determine the optimal voltage value to ensure effective performance. To analyze the impact of electrode voltage on the electric field distribution, simulations were conducted. According to the simulation model established in Sect. 8.4.1, the field strength distribution under different voltages was calculated. Figure 8.8 illustrates the field strength distribution at various voltages.

Experimental results from the literature indicate that, with a constant flow rate, the charge-to-mass ratio of the electrode varies with voltage. As the voltage increases, the charge-to-mass ratio initially rises and reaches a peak before decreasing. The decrease in charge-to-mass ratio in the second phase is due to ionization of the working medium at high voltages, disrupting the balance of normal charging and leading to unstable conditions and corona discharge. From the electric field strength curves in Fig. 8.8, it can be observed that when the voltage increases from 2 to 14 kV, the field strength at the nozzle increases to approximately four times its original value. This indicates a proportional relationship between electric field strength and voltage. As the absolute value of the voltage increases, the field strength at the electrodes

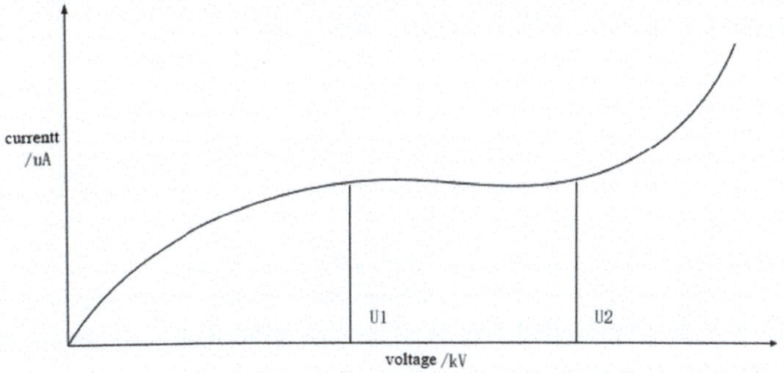

Fig. 8.7 Gas conductive volt-ampere characteristic curve

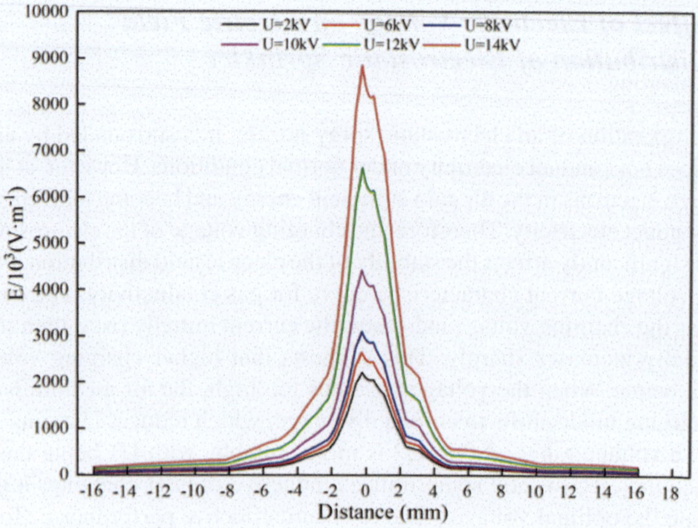

Fig. 8.8 Electric field distribution at different voltage electrodes of electrostatic nozzle

also increases, and the rate of increase becomes more pronounced. This is due to the stronger interaction between charges at higher voltages, leading to a significant increase in electric field strength. However, excessively high voltages pose risks such as insulation breakdown between electrodes. Insulation breakdown occurs when the medium (e.g., air) between the electrodes loses its insulating properties at high field strengths, leading to discharge phenomena. This not only destabilizes the electric field but may also damage equipment and pose safety hazards. Considering these factors, the optimal output voltage for high-voltage electrostatic generators during electrostatic spraying operations is around 10 kV.

8.5.3 Effect of Different Ring Sizes on the Electric Field Distribution in Electrostatic Spraying

Based on the theoretical analysis in Sect. 8.5.2, it is evident that the dimensions of the ring electrode in an electrostatic nozzle (diameter and thickness) have a notable impact on the electric field distribution. To thoroughly observe and analyze this effect, different cross-sections of the ring's charged region were selected for simulation. Specifically, the cross-sections at y = 5 mm and y = 10 mm were chosen, with the ring electrode thickness set at 4 mm, while other parameters were kept constant. As shown in Fig. 8.9, the radial and axial electric field distributions are compared for ring diameters of 32 mm, 36 mm, and 40 mm, all with the same thickness. It is clear from the figure that the axial field strength is consistently greater than the radial field strength for the same ring structure. Figure 8.9a illustrates that the electrostatic

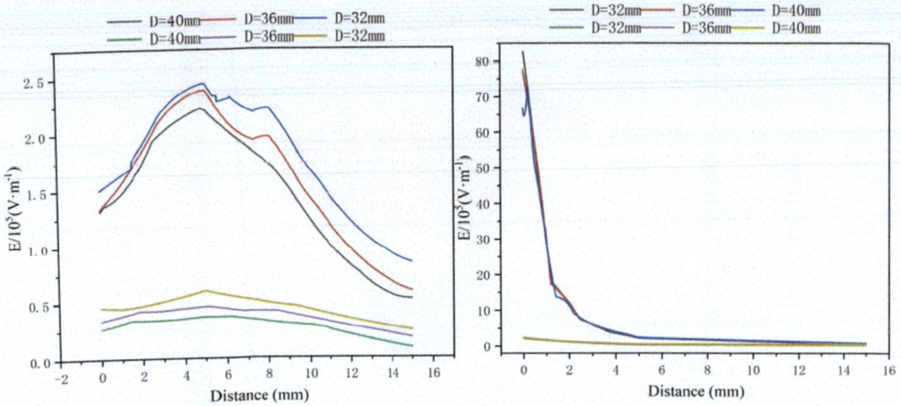

Fig. 8.9 The radial and axial field strength distribution of the nozzle under different annular electrode diameters

spray field expands radially, reaching a peak before diminishing. Furthermore, the difference in field strength at the axis of symmetry is minimal across rings with different diameters. With the same electrode thickness, a smaller electrode diameter results in a relatively higher electric field strength. The curves in the figure show an increasing gap as the radial distance from the central axis grows, indicating that the field strength differences between rings with varying diameters are more pronounced near the electrode surface, while the effect on the central axis is minimal.

Figure 8.9b shows that the axial field strength distribution for different ring diameters is nearly identical, implying that the dimensions of the ring electrode have little to no effect on the axial field strength. In other words, the size of the ring electrode has an insignificant impact on the axial field strength of the electrostatic nozzle. However, as the cross-sectional distance increases, it becomes apparent that the ring structure significantly affects the radial field strength. To further analyze the interaction between the outer diameter and thickness of the electrode, simulations were conducted on different combinations. The inner diameters of 20 mm, 24 mm, 28 mm, and 32 mm were paired with electrode thicknesses of 2 mm, 3 mm, 4 mm, and 5 mm, respectively, yielding 16 combinations in total. The maximum values of the electric field strength Mag_E for each combination are shown in Fig. 8.10. The data reveal that reducing the inner diameter or increasing the thickness of the ring electrode significantly affects the electrostatic field strength distribution. Therefore, when designing the dimensions of the ring electrode for an electrostatic nozzle, it is recommended to use a smaller inner diameter and a larger electrode thickness to enhance the droplet charging capability.

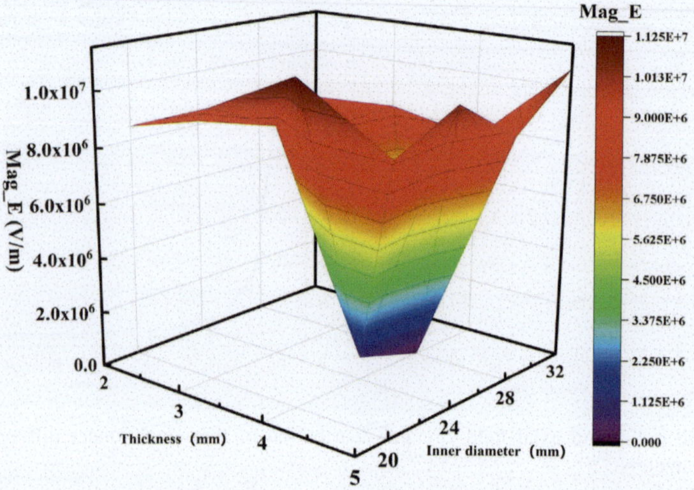

Fig. 8.10 Electric field strength values corresponding to different electric ring sizes

8.5.4 Effect of the Distance Between the Nozzle and the Electric Ring on the Electric Field Strength

In addition to the structural parameters of the ring electrode, the installation distance between the nozzle and the ring electrode also influences the distribution of the electrostatic field. In this electrostatic nozzle, the nozzle and the ring electrode form an inductive charging device, with the nozzle positioned behind the ring electrode. The distance between the two, denoted as H, affects the field distribution, as illustrated in the layout diagram. To explore how this installation distance affects the electric field strength, a simulation was conducted with the coordinate difference between the nozzle tip and the upper surface of the ring electrode as a variable. Five different electrode distances H were modeled: 0 mm, 2 mm, 5 mm, 10 mm, and 20 mm. The ring electrode had an inner diameter of 32 mm, an outer diameter of 42 mm, and a thickness of 5 mm. The charging voltage was consistently set to 10 kV. Figure 8.11 illustrates the simulation model, which includes the cross-section, ring electrode, and nozzle.

As shown in Fig. 8.12, the simulation was conducted by varying the distance H between the nozzle and the ring electrode, and the electric field distribution across the nozzle's cross-section at different distances was extracted. The electric field intensity contour maps visually demonstrate the distribution of the electric field across different regions of the plane, with colors typically representing varying field intensities. For example, red or yellow areas indicate regions of higher field strength, while green may represent areas with moderate field strength. When the electric field strength is concentrated in specific regions of the nozzle plane, this typically suggests stronger field intensities in those areas, which could enhance the charging effect on the

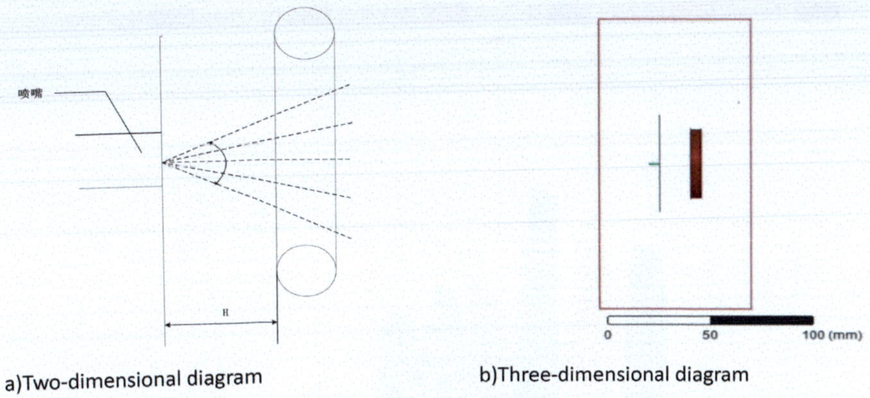

Fig. 8.11 Distance model between nozzle and electric ring

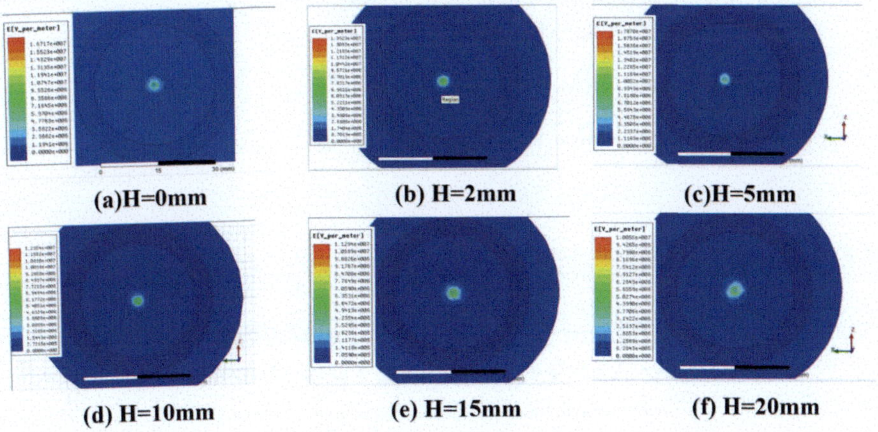

Fig. 8.12 Simulation results of different spacing H on the same plane

droplets. To further clarify the effective limit of the electric field interaction between the nozzle and the ring electrode, the variation in electric field strength along the ring electrode's axis at different distances was plotted. From the vertical axis in Fig. 8.13, it can be observed that configurations with smaller distances between the nozzle and the ring electrode produce higher electric field strengths. Therefore, in nozzle design, selecting a distance of less than 5 mm can result in a stronger electric field, which is beneficial for improving the droplet charging performance of the nozzle.

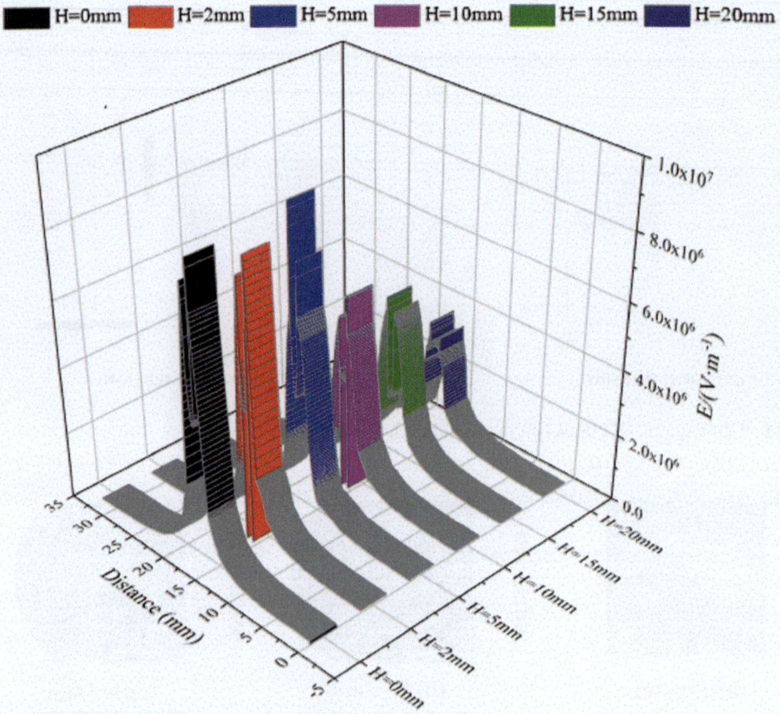

Fig. 8.13 Electric field distribution at different electrode spacings

8.5.5 Effect of Different Spacing Between Multiple Electrostatic Nozzles on Electric Field Distribution

In electrostatic spraying, the induced electric field is stronger in areas closer to the ring electrode. To study the electrostatic effects of multiple nozzles, we assume that the point charges of the ring electrodes for the multiple nozzles are represented by T1, T2, and T3. The electric field intensities at a specific point T in space, generated by each ring electrode, are denoted as TE1′, TE2′ and TE3′, respectively. The combined electric field intensity at point T is represented as TE, and the simplified model of the multi-nozzle electric field is shown in Fig. 8.14. Since the electric fields at point T are additive, the total electric field is $TE' = TE1' + TE2' + TE3'$, and $|TE'| > |TE1'|$, indicating that the electric field intensity of multiple nozzles is theoretically higher than that of a single nozzle. In the simulation model, we used a ring electrode with an inner diameter of 16 mm, an outer diameter of 42 mm, a thickness of 5 mm, and a voltage of 10 kV, with boundary dimensions of 500 mm × 2000 mm × 500 mm and a nozzle-ring electrode distance of 0 mm. The distances between multiple nozzle electrodes were set to 200 mm, 300 mm, and 400 mm for comparison. The electric field intensity contour maps for both single and multiple nozzles, as well as the average electric field values along the vertical spray rod for each cross-section, are

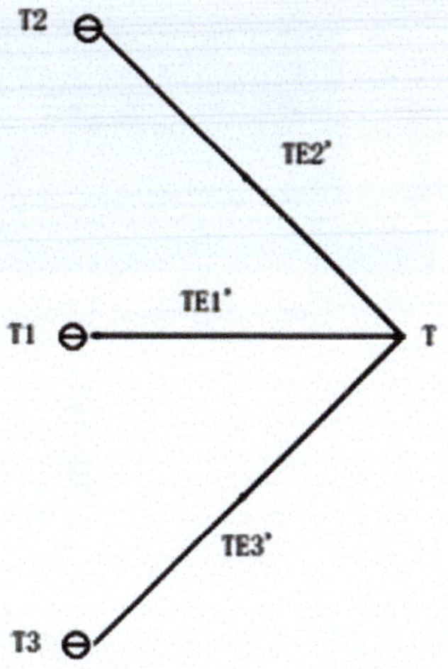

Fig. 8.14 Simplified electric field model of vertical spray boom multi-electric ring nozzle

shown in Figs. 8.15 and 8.16. From these figures, it is evident that the electric field intensity is higher near the nozzle electrodes, and the field intensity of a single nozzle is lower than that of multiple nozzles. Since the electric field contour maps do not clearly show the variation in field strength, the electric field intensity along the vertical spray rod indicates that the average field strength of multiple nozzles is higher than that of a single nozzle. This is due to the overlapping of electric fields between the nozzles, leading to a higher electric field intensity in the overlapping regions. This phenomenon enhances the effect of electrostatic spraying in the working space, improving droplet attachment across the entire area. Additionally, the field strength decreases more rapidly for a single nozzle, leading to lower uniformity, whereas the field strength for multiple nozzles decays more slowly, resulting in better uniformity. The smaller the distance between nozzles, the more pronounced this effect. Considering the growth characteristics of greenhouse tomatoes and practical working conditions, although a smaller nozzle distance generates a stronger electric field, it also increases the likelihood of corona discharge. Furthermore, increasing the number of nozzles may conflict with water-saving requirements. Therefore, a nozzle spacing of 300 mm is the most suitable for this working environment.

The geometric parameters of the vertical spray rod electrostatic sprayer are closely related to the distribution of the electrostatic field, which influences droplet charging and the sprayer's operational performance. This chapter uses Ansoft Maxwell electrostatic field simulation software to analyze various parameter configurations, including electrostatic voltage, ring electrode structure, and nozzle distribution. The simulation results provide a foundation for future optimization experiments in electrostatic spraying, and the following conclusions were drawn (1) the electrostatic field of the vertical spray rod with multiple nozzles is symmetrically distributed. The field strength increases closer to the nozzle, first decreases along the spraying direction, then increases, and finally decreases again after moving away from the

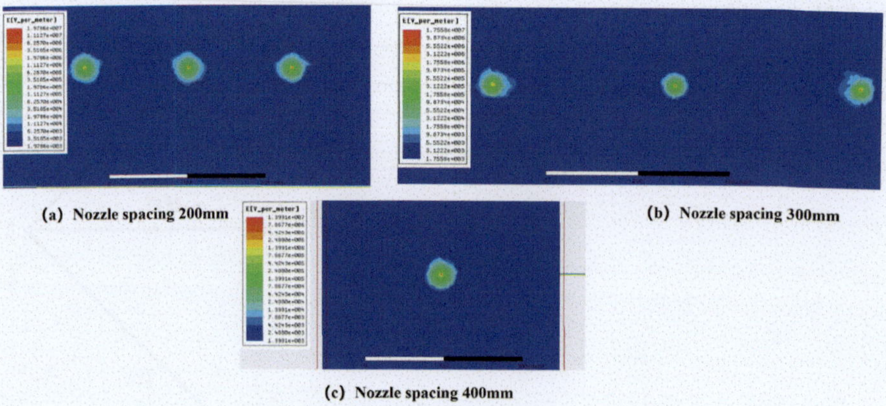

Fig. 8.15 Simulation results of different nozzle spacing on the same plane

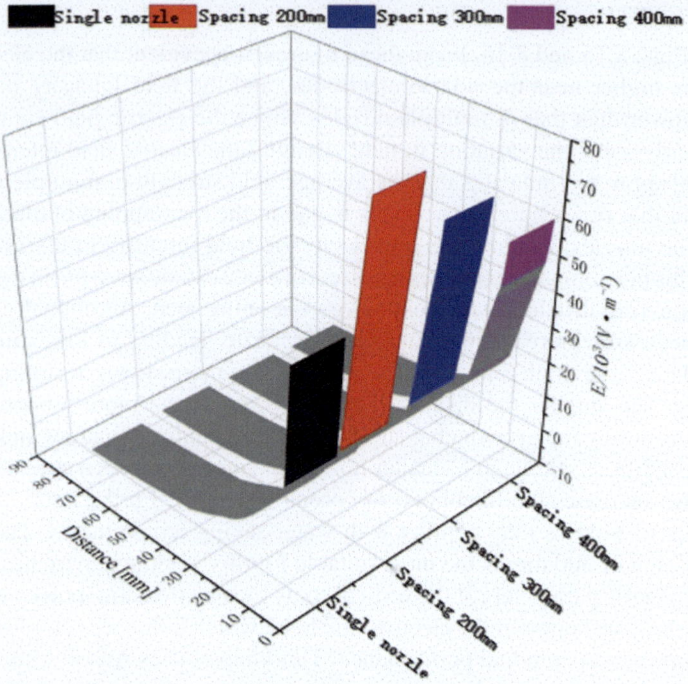

Fig. 8.16 Electric field distribution at different nozzle spacings

electrode. This indicates that the droplet charging zone is relatively small, and high-quality atomization is required for nozzle selection. (2) The strength of the electrostatic field in inductive charging electrostatic spraying increases with charging voltage. However, beyond a certain point, the charging voltage must be controlled to avoid corona discharge, which can affect spray quality. Simulation results indicate that a generator output voltage of 10 kV is optimal. (3) The diameter of the ring electrode significantly affects the radial electric field strength. Smaller ring diameters lead to higher radial field strengths, increasing the radial electrostatic force on droplets, expanding the spray cone angle, and enlarging the charged droplet area. This improves the operational effectiveness when used with a fan. Meanwhile, axial field strength remains nearly constant regardless of changes in the ring diameter. Based on wind tunnel parameters, the optimal ring electrode structure is an inner diameter of 32 mm and a thickness of 5 mm. (4) Simulation results indicate that smaller distances between the nozzle and the ring electrode produce higher electric field strengths. For this sprayer's ring electrode design, a distance of less than 5 mm is ideal for maximizing electric field strength and enhancing droplet charging. (5) Simulations of multiple electrostatic nozzles show that the electric field intensity with multiple nozzles is higher than that of a single nozzle under the same parameters, and the field intensity increases more significantly as the nozzle spacing decreases. However, with a spacing of 400 mm, the field strength increases only slightly. Considering the design parameters of the spray boom, a spacing of 300 mm is optimal for this application.

8.6 Single Factor Experiment and Analysis

8.6.1 Electrostatic and Non-electrostatic Comparison Test

Based on the experimental preparation and plan, the connection status of each component of the electrostatic spray machine was checked, and a greenhouse tomato experiment was conducted. Water-sensitive paper was arranged as required, with a spraying distance of 40 cm and a spraying pressure of 0.6 MPa. After the spray machine reached a stable spraying condition, the self-propelled button was activated, allowing the machine to operate at a speed of 0.5 m/s. Electrostatic voltages of 0 and 10 kV were compared. The above experiment was repeated, and the attachment of droplets to the front and back sides of the tomato plants was measured according to the measurement method, as shown in Fig. 8.17.

The results of the tomato plant leaf attachment test under electrostatic conditions indicate that when the spray machine operates with fixed parameters, the attachment on the leaf surface's edges is good. The electrostatic effect significantly improves leaf attachment. The average droplet coverage rates were 84.50% on the front and 44.24% on the back under electrostatic conditions, compared to 70.33% on the front and 21.93% on the back without electrostatic effects. When the electrostatic effect

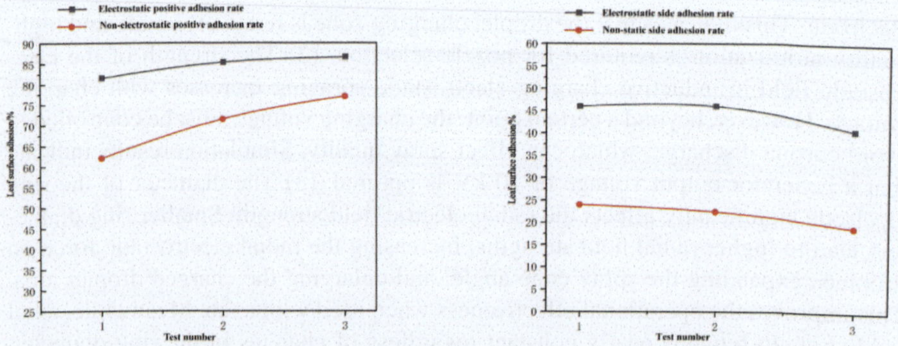

Fig. 8.17 Static and non-static attachment to the front and back of plant leaves

was applied, the average attachment on the front and back of the plant leaves increased by 14.17% and 22.31%, respectively. Analysis shows that the electrostatic effect has a notable impact on leaf attachment, enhancing the adhesion of droplets.

8.6.2 Single Factor Test of Electrostatic Spray Distance

Based on the experimental preparation and plan, the connection status of each component of the electrostatic spray machine was inspected. A greenhouse tomato experiment was conducted by arranging water-sensitive paper as required. The remote control switch and electromagnetic valve were activated, the water pump pressure was adjusted to 0.6 MPa, and the electrostatic generator voltage was set to 6 kV. After the spray machine reached a stable spraying condition, the automatic walking button was pressed, causing the machine to operate at a speed of 0.5 m/s. The experiment was repeated at different spraying distances, and after completing the work, the attachment of droplets to the front and back sides of the tomato plants was measured according to the method, as shown in Fig. 8.18.

According to the results of the single-factor experiment on electrostatic spraying distance, the attachment on the front and back sides of the leaves showed complex variations. Within the range of 25–50 cm, the coverage on the front of the leaves initially increased but then decreased as the distance increased. Conversely, the coverage on the back of the leaves gradually decreased with increasing spraying distance. The reason for this phenomenon is that, for the front attachment, when the spraying distance is too short, the droplets have not had time to disperse due to the nozzle's designed spray radius, resulting in reduced droplet attachment space on the leaf. Additionally, this can cause excessive droplet concentration on the front of the leaf, negatively affecting the spraying effectiveness. For the back of the leaves, the charge on the droplets decreases as the distance increases, eventually approaching zero after a certain distance. At this point, the droplets are more influenced by gravity

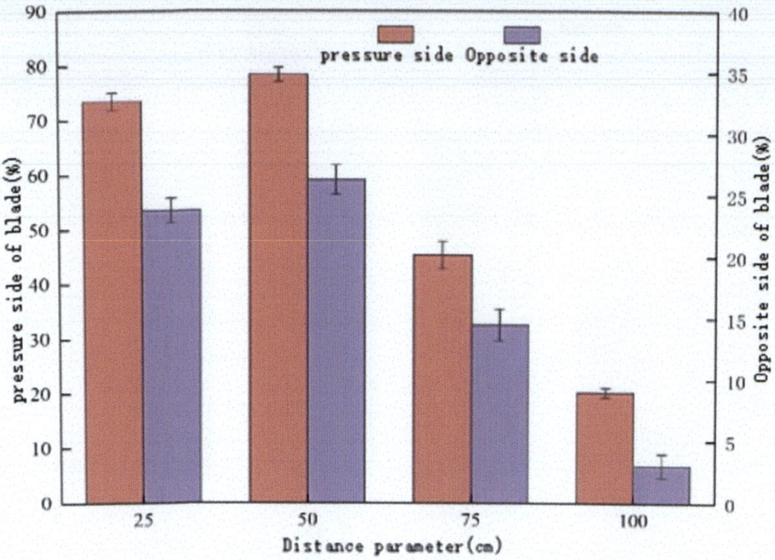

Fig. 8.18 Single factor relationship between spray distance and leaf front and back attachment

than by the electrostatic field, which hinders the formation of the enveloping effect. These experimental results are consistent with the simulation results of the axial electric field distribution in the electrostatic field.

8.6.3 *Single Factor Test of Working Pressure of Electrostatic Spray*

Based on the experimental preparation and plan, the connection status of each component of the electrostatic spray machine was inspected. A greenhouse tomato experiment was conducted by arranging water-sensitive paper as required. The electrostatic generator voltage was set to 6 kV with a spraying distance of 50 cm, and the spraying pressure was determined. After the spray machine reached a stable spraying condition, the automatic walking button was activated, allowing the machine to operate at a speed of 0.5 m/s. The experiment was repeated at different working pressures, and after completion, the attachment of droplets to the front and back sides of the tomato plants was measured according to the method, as shown in Fig. 8.19. According to the results of the single-factor experiment on electrostatic working pressure, the droplet coverage on the front and back of the leaves initially increased with spraying pressure but then decreased as the pressure continued to rise. Based on the principle of droplet atomization under pressure, it is known that when the spraying pressure is low, the droplets are not fully atomized, resulting in larger droplet sizes. At this stage, the droplets experience greater gravitational force compared to the effects

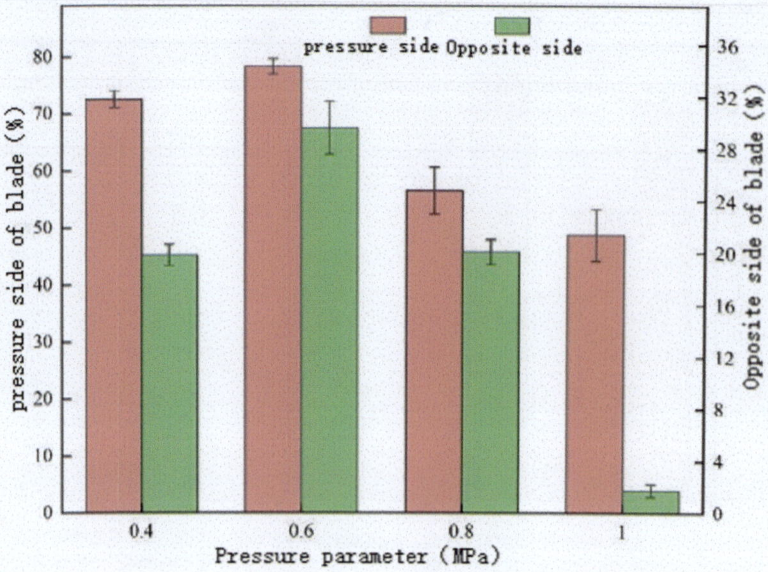

Fig. 8.19 Single factor relationship between spray working pressure and blade front and back attachment

of wind and the electrostatic field, making it difficult for them to deposit on the leaf surface. As the spraying pressure increases, the atomization effect improves, resulting in smaller droplet sizes. This enhances the effects of wind and electrostatic forces, increasing the attachment on both the front and back of the leaf surfaces. However, when the pressure becomes too high, the droplets become very fine with minimal gravitational force, making them prone to drifting under the influence of wind. Additionally, under the same charge conditions, like-charged droplets repel each other, which further impedes their attachment to the target plant leaves and affects the quality of spraying.

8.6.4 Electrostatic Voltage Single Factor Test of Electrostatic Spray

Based on the experimental preparation and plan, the connection status of each component of the electrostatic spray machine was inspected. A greenhouse tomato experiment was conducted, arranging water-sensitive paper as required. The remote control switch and electromagnetic valve were activated, the water pump pressure was set to 0.6 MPa, and the spraying distance was adjusted to 50 cm. The high-voltage electrostatic generator voltage was varied according to the experimental parameters. After the spray machine reached a stable spraying condition, the automatic walking button

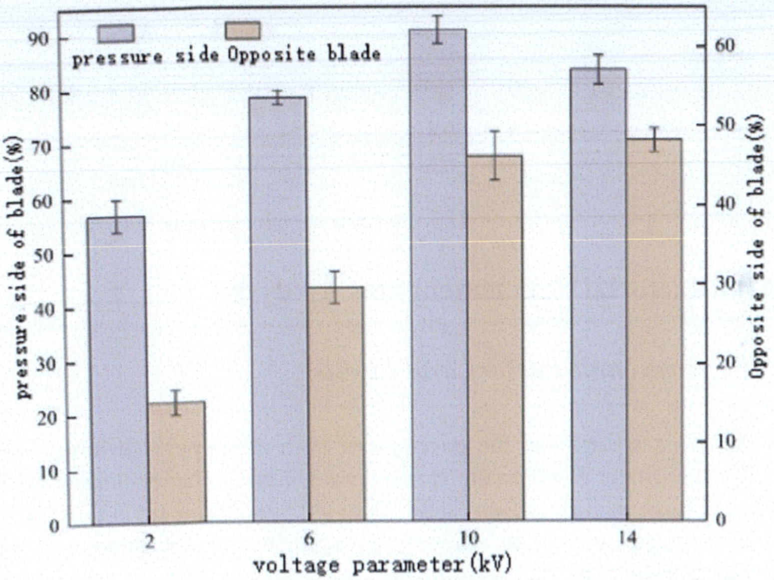

Fig. 8.20 Single factor relationship between electrostatic voltage and blade front and back attachment

was pressed, allowing the machine to operate at a speed of 0.5 m/s. The experiment was repeated at different working voltages, and after completion, the attachment of droplets to the front and back sides of the tomato plants was measured according to the method, as shown in Fig. 8.20. According to the results of the single-factor experiment on electrostatic working voltage, the droplet attachment on the front of the leaves initially increased with the increase in electrostatic voltage, reached a peak, and then decreased. In contrast, the attachment on the back of the leaves gradually increased. Analysis reveals that under the influence of electrostatic force, an electrostatic field forms between the droplets and the plants. As the electrostatic voltage increases, the charge on the droplets also increases, enhancing the attachment on the front and back of the leaves. However, when the charging voltage reaches 14 kV, there is a downward trend in the coverage on the front of the leaves. During the experiment, it was observed that at excessively high charging voltages, droplets accumulated on the electrostatic nozzle's electrode ring and showed a tendency to attract charged droplets. This might be due to the opposite polarity of the droplets accumulated on the electrode ring compared to the droplets being sprayed out, resulting in reverse ionization. This phenomenon reduces the charge on the droplets and causes swirling of the droplets around the plant. The effect becomes more pronounced with higher voltage. Additionally, at high electrostatic voltages, a corona phenomenon is likely to form on the electrode ring, which is consistent with the simulation results of the electrostatic field distribution affected by the electrode voltage. Ultimately, this will impact the coverage rate of the water-sensitive paper on the leaves.

Table 8.2 Test level table

Level	Factor		
	Static voltage level D (kV)	Work pressure P_r (MPa)	Target distance L (cm)
−1	6	0.4	25
0	10	0.6	50
1	14	0.8	75

8.7 Multifactorial Experiment and Analysis

8.7.1 Experimental Factors and Levels

In a single-factor experiment, the coverage of both the upper and lower surfaces of tomato leaves under electrostatic spraying was used as the evaluation criterion. The effects of three factors—electrostatic spraying pressure, spraying distance, and electrostatic voltage—on leaf coverage were investigated to determine their optimal working ranges. The working pressure range was selected from 0.4 to 0.8 MPa, the spraying distance from 25 to 75 cm, and the electrostatic voltage from 6 to 14 kV. To explore the interaction effects of these factors on spray performance and identify the optimal parameter combination, a three-factor, three-level quadratic orthogonal rotational center composite design can be employed. This approach allows for the establishment of a mathematical model between the influencing factors and response variables, describing the quantitative relationship between them. The results are optimized to ultimately determine the best parameter combination for electrostatic spraying operations in crop protection. The experimental levels for each factor are shown in Table 8.2.

8.7.2 Analysis of Response Surface Model for Electrostatic Spraying

The response variable for the experiment was defined as the ratio of droplet area to the total area on water-sensitive paper, expressed as a percentage. The experiment design and analysis of the sprayer's operation were conducted using Design-Expert 13.0 software. The droplet coverage test results are shown in Table 8.3, and a subset of the water-sensitive paper test results and analyses is illustrated in Fig. 8.21.

A total of 17 sets of data were collected and presented in Table 8.3. Based on this data, Design-Expert 13.0 was used to fit the experimental data, and a regression equation describing the effects of target distance, electrostatic voltage level, and working pressure on droplet adhesion rates on both sides of the leaves during sprayer operation was obtained as follows.

Table 8.3 Droplet coverage test results

Test no	Experimental factors and levels			Response value	
	Static voltage level D (kV)	Work pressure Pr (MPa)	Target distance L (cm)	Droplet adhesion rate on the front of leaves Z (%)	Droplet adhesion rate on the back of leaves F (%)
1	6	0.4	50	72.54	19.83
2	6	0.8	50	56.58	20.07
3	10	0.4	25	85.36	21.82
4	14	0.8	50	63.95	50.46
5	10	0.6	50	79.45	31.66
6	14	0.4	50	75.45	28.85
7	10	0.6	50	85.74	35.56
8	14	0.6	25	89.56	45.24
9	10	0.4	75	43.85	7.16
10	10	0.6	50	80.09	30.11
11	14	0.6	75	41.45	10.28
12	6	0.6	25	73.49	4.79
13	10	0.6	50	84.24	32.65
14	10	0.6	50	83.14	30.17
15	10	0.8	25	82.48	40.62
16	6	0.6	75	45.31	5.42
17	10	0.8	75	44.92	7.42

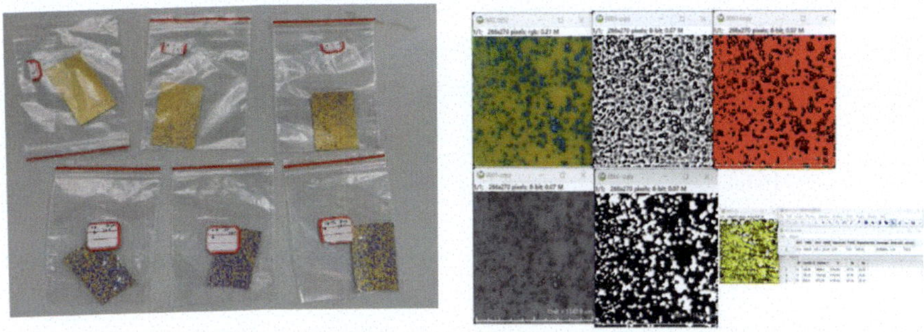

Fig. 8.21 Partial water-sensitive paper test results and analysis

Where D represents the electrostatic voltage level, kV; Pr represents the working pressure, MPa; L represents the target distance, cm. The regression model analysis of the droplet adhesion rates on both sides of the leaves during electrostatic spraying is presented in Tables 8.4 and 8.5.

Table 8.4 Regression model analysis of droplet adhesion rate on the front of electrostatic sprayer blades

Variable	Sum of squares	Degrees of Freedom	Mean square	F-value	*P*-value	Significance
Model	4478.08	9	497.56	30.90	<0.0001	**
D	63.23	1	63.23	3.93	0.0880	
P_r	107.09	1	107.09	6.65	0.0365	*
L	3017.09	1	3017.09	187.38	<0.0001	**
DP_r	4.97	1	4.97	0.3088	0.5957	
DL	99.30	1	99.30	6.17	0.0420	*
P_rL	3.90	1	3.90	0.2422	0.6376	
D^2	307.87	1	307.87	19.12	0.0033	**
P_r^2	197.63	1	197.63	12.27	0.0099	**
L^2	559.61	1	559.61	34.75	0.0006	**
Residual	112.71	7	16.10			
Lack of Fit	83.67	3	27.89	3.84	0.1132	
Error	83.67	4	7.26			
Sum	4590.79	16				

The more asterisks there are, the higher the significance

Table 8.5 Regression model analysis of droplet attachment rate on the back of electrostatic sprayer blades

Variable	Sum of squares	Degrees of Freedom	Mean square	F-value	*P*-value	Significance
Model	3232.05	9	359.12	52.01	<0.0001	**
D	897.18	1	897.18	129.93	<0.0001	**
P_r	209.20	1	209.20	30.30	0.0009	**
L	844.40	1	844.40	122.29	<0.0001	**
DP_r	114.17	1	114.17	16.53	0.0048	**
DL	316.66	1	316.66	45.86	0.0003	**
P_rL	85.93	1	85.93	12.45	0.0096	**
D^2	26.84	1	26.84	3.89	0.0893	
P_r^2	0.3727	1	0.3727	0.0540	0.8229	
L^2	719.54	1	719.54	104.21	<0.0001	**
Residual	48.34	7	6.91			
Lack of Fit	28.21	3	9.40	1.87	0.2757	
Error	20.13	4	5.03			
Sum	3280.38	16				

The more asterisks there are, the higher the significance

According to the regression model analysis for droplet adhesion rates on the upper leaf surface, each influencing factor had a certain effect on the response variable. The larger the *p*-value, the less significant the effect. The analysis indicates that the factors L, D^2, P^2, and L^2 had a highly significant impact ($P < 0.01$) on droplet adhesion rates, while P and DL had a significant impact ($0.01 < P < 0.05$). However, D, DP, and PL were not significant ($P > 0.05$). Similarly, for the regression model of droplet adhesion rates on the lower leaf surface, all influencing factors had some effect on the response variable. Again, larger *p*-values indicate less significant effects. The analysis shows that PL, D^2, P^2, and L^2 had a highly significant impact ($P < 0.01$) on droplet adhesion rates, while D and L had a significant impact ($0.01 < P < 0.05$). However, P, DP, and DL were not significant ($P > 0.05$). The *p*-values for the models calculated in Tables 8.4 and 8.5 were both < 0.0001, meeting the criterion for high significance ($P < 0.01$). The *p*-values for the lack of fit in the models were greater than 0.05, indicating the models' fit was appropriate. The determination coefficients (R2) for models Z and Y were 97.54% and 98.53%, respectively, indicating that over 97% of the variation in the response values could be explained by these two models.

8.7.3 *Analysis of Factors Affecting Electrostatic Spray Response Index*

8.7.3.1 Effects of Various Factors on the Front Droplet Adhesion Rate of Electrostatic Operation

Based on the significance of the *p*-values in the table, the order of influence of the factors on the droplet adhesion rate on the upper surface of leaves in electrostatic spraying is as follows: target distance L > working pressure Pr > electrostatic voltage level D. The response surface curve in Fig. 8.22 illustrates the interaction between the two factors, electrostatic voltage level D and working pressure Pr, on the droplet adhesion rate Z on the upper surface when the target distance is at the central level ($L = 50$). When $L = 50$, the droplet adhesion rate Z increases as D and Pr increase, but after reaching a certain level, it stops increasing and may even decrease. At lower working pressures $Pr = 0.8$, the upper surface droplet adhesion decreases as the voltage increases. This phenomenon occurs because the higher spray pressure excessively atomizes the droplets, and the effect of wind transport becomes more significant than the electrostatic force, which may lead to noticeable droplet repulsion, negatively impacting adhesion.

Figure 8.23 shows the response surface curve depicting the interaction between D and L on the droplet adhesion rate Z when the working pressure is at the central level $P = 0.6$. When $P = 0.6$, the droplet adhesion rate Z increases slowly as D increases. However, as the spraying distance increases, the droplet adhesion rate Z first increases and then decreases.

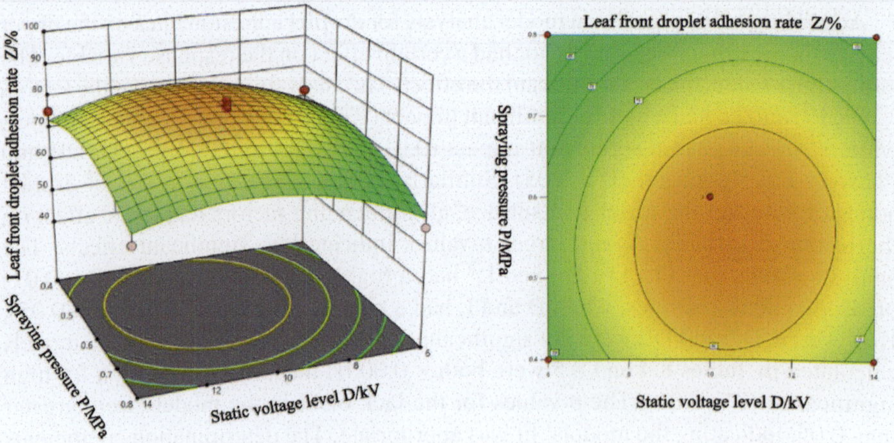

Fig. 8.22 Interaction response surface and contour lines of electrostatic voltage and working pressure

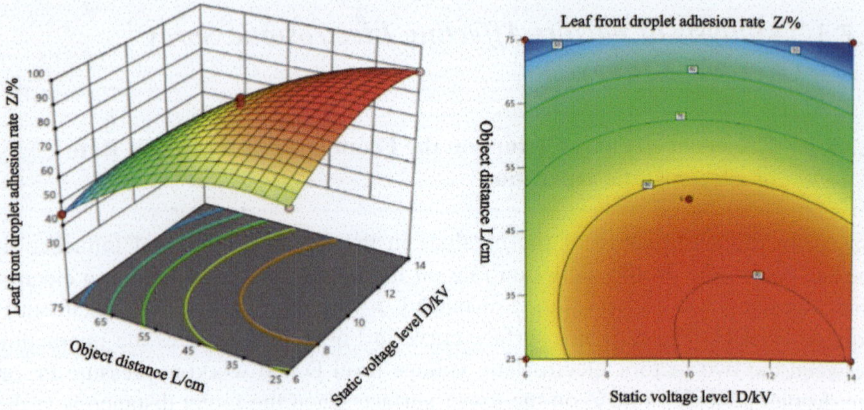

Fig. 8.23 Interaction response surface and contour lines of electrostatic voltage and target distance

In Fig. 8.24, the response surface curve shows the interaction between Pr and L on the droplet adhesion rate Z when the electrostatic voltage is at the central level D = 10. When D = 10, the droplet adhesion rate Z initially increases as Pr and L increase but then begins to decrease.

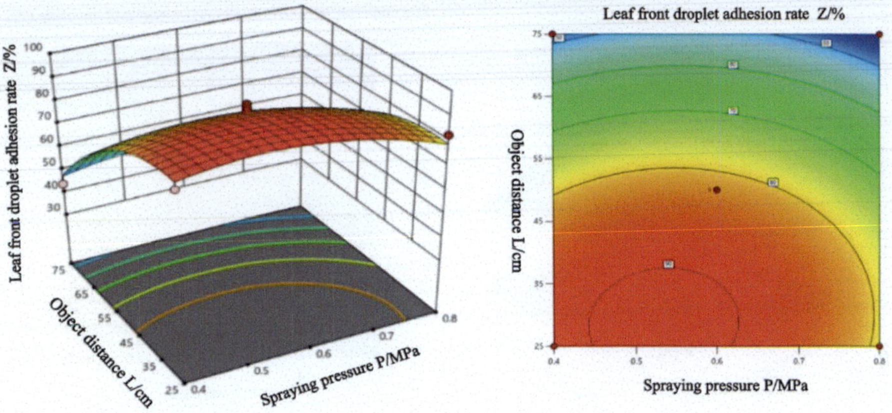

Fig. 8.24 Interaction response surface and contour lines of working pressure and target distance

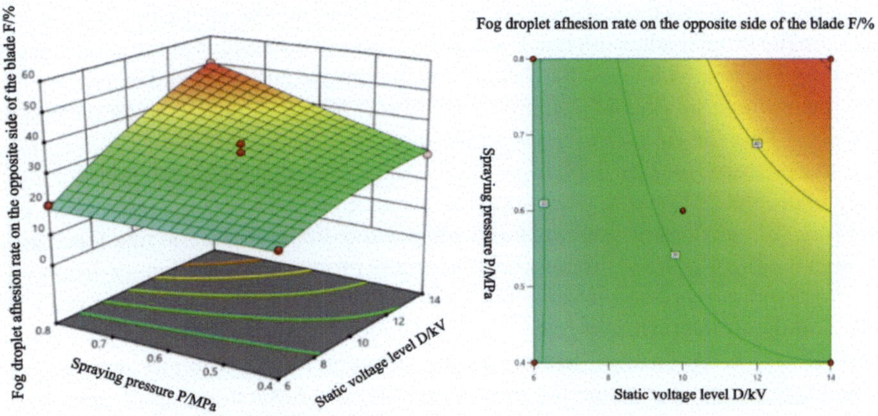

Fig. 8.25 Interaction response surface and contour lines between working pressure and electrostatic voltage

8.7.3.2 Effects of Various Factors on the Droplet Adhesion Rate on the Reverse Side of Electrostatic Operation

Based on the significance of the *p*-values in the table, the order of influence of the factors on the droplet adhesion rate on the underside of leaves during electrostatic spraying is as follows: electrostatic voltage level D > target distance L > working pressure Pr. The response surface curve in Fig. 8.25 illustrates the interaction between the electrostatic voltage level D and working pressure Pr on the droplet adhesion rate F on the underside of leaves when the target distance is at the central level (L = 50). When L = 50, the droplet adhesion rate F increases with increasing D and Pr but eventually plateaus or decreases. This decline is likely due to the overly fine droplets being increasingly affected by environmental factors when the pressure and charging voltage become too high.

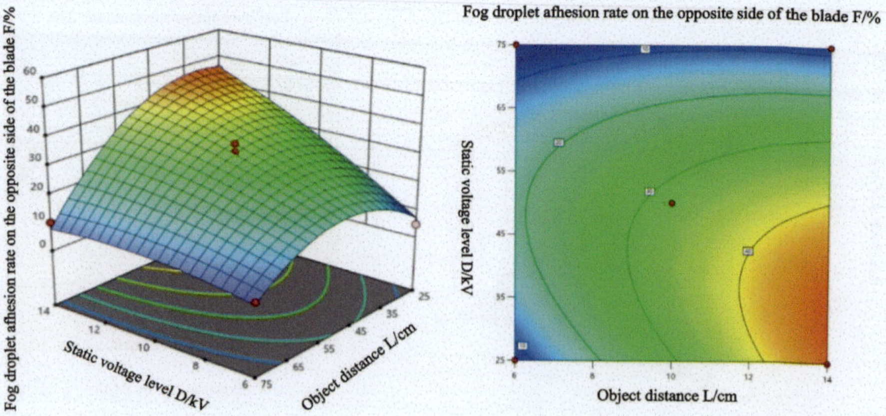

Fig. 8.26 Target distance and electrostatic voltage interaction response surface and contour lines

Figure 8.26 shows the response surface curve depicting the interaction between D and L on the droplet adhesion rate F when the working pressure is at the central level (P = 0.6). The graph indicates that decreasing the spraying distance and increasing the charging voltage significantly improve the droplet adhesion rate F on the underside of leaves. This improvement is due to the fact that charged droplets have a certain adhesive property, making them more likely to adhere compared to uncharged or less-charged droplets. However, if the spraying distance is relatively long, increasing the charging voltage does not significantly enhance the droplet adhesion rate F because the attenuation of charged droplets becomes more pronounced over longer distances.

The response surface curve in Fig. 8.27 illustrates the interaction between Pr and L on the droplet adhesion rate F when the electrostatic voltage is at the central level (D = 10). When D = 10, the droplet adhesion rate F generally increases as Pr and L decrease. The reduction in spraying distance has a more significant effect on the adhesion rate F than increasing the working pressure. This is because, with appropriate airflow, shorter spraying distances reduce the attenuation of charged droplets, resulting in better deposition.

8.7.4 Electrostatic Spraying Operation Performance Verification Test

Based on the results of the multi-factor experiment analysis and the interaction effects between the three influencing factors, the reliability and safety of the spraying operation were considered. Using the optimization function in Design-Expert software, the levels of the influencing factors for the spraying operation were optimized as a target function. As a result, the optimized operational parameters for the facility's electrostatic sprayer were identified as: an electrostatic voltage level of +10 kV, a working

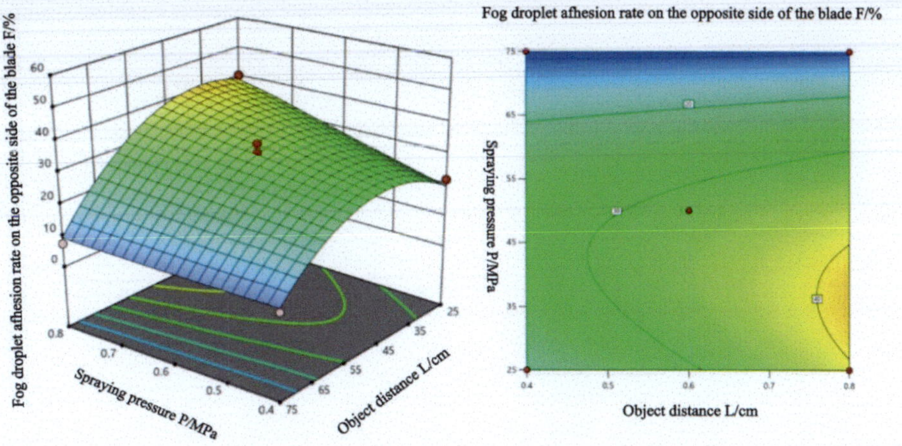

Fig. 8.27 Target distance and working pressure interactive response surface and contour lines

Table 8.6 Field test results

Response value	Predicted value (%)	Test value (%)	Relative error (%)
Z	86.15	84.11	2.37
		87.38	1.4
		79.08	8.2
F	37.55	33.88	9.77
		35.12	6.47
		39.71	5.75

pressure of 0.7 MPa, and a spraying distance of 35 cm. Under these optimal conditions, the predicted adhesion rates on the upper and lower surfaces of the greenhouse tomato leaves were 86.15% and 37.55%, respectively.

The optimized parameters were tested three times in the experimental field, and the average values were obtained, as shown in Table 8.6. The relative error between the predicted and actual values for the upper surface adhesion of the tomato leaves was less than 9%, and for the lower surface adhesion, less than 10%. The highest adhesion rates for the upper and lower surfaces were 87.38% and 39.71%, respectively, with an average adhesion rate of 83.52% for the upper surface and 36.24% for the lower surface across the three tests. These results indicate that the errors between the predicted and actual values were minimal, suggesting that the response surface model for electrostatic spraying has practical significance for guiding real-world applications. When the electrostatic voltage level was set to +10 kV, the working pressure to 0.7 MPa, and the spraying distance to 35 cm, the overall optimal adhesion rates for the upper and lower surfaces of the greenhouse tomato leaves were achieved, with values of 83.52% and 36.24%, respectively.

Chapter 9
Design and Experiment of Intelligent Spraying Machine for Desert Gobi Facilities

Long Wang, Xufeng Wang, Can Hu, Yi Zhu, Guansan Zhu, and Wanjia Hua

9.1 Introduction

The cultivation of tomatoes in solar greenhouses in southern Xinjiang is dense, with tall plants and abundant foliage. The temperature and humidity within the greenhouses are high, and the plants grow in a relatively enclosed environment. Consequently, pest and disease problems with tomato crops are becoming increasingly severe. The existing spraying machinery does not effectively target the undersides of leaves where pests and diseases often hide, which negatively impacts tomato growth and yield, hindering the promotion of the tomato industry in southern Xinjiang's greenhouse facilities.

In foreign countries, specialized pest control machinery for greenhouses has been developed by incorporating advanced technologies such as targeted positioning and

L. Wang · X. Wang (✉) · C. Hu · Y. Zhu · G. Zhu
College of Mechanical and Electrical Engineering, Tarim University, Alar 843300, China
e-mail: wxfwyq@126.com

Modern Agricultural Engineering Key Laboratory at Universities of Education Department of Xinjiang Uygur Autonomous Region, Alar 843300, China

Xinjiang Production and Construction Corps (XPCC) Key Laboratory of Utilization and Equipment of Special Agricultural and Forestry Products in Southern Xinjiang, Alar 843300, China

L. Wang
e-mail: 120140002@taru.edu.cn

W. Hua
Key Laboratory of Smart Agriculture System Integration, Ministry of Education, Beijing 100083, China
e-mail: s20223071396@cau.edu.cn

Key Laboratory of Agricultural Information Acquisition Technology, Ministry of Agriculture and Rural Affairs, Beijing 100083, China

C. Hu et al. (eds.), *New Technologies Applied in Gobi Desert Facility Agriculture*, Smart Agriculture 11, https://doi.org/10.1007/978-981-96-1031-0_9

electrostatic spraying, significantly improving the effectiveness of pest control operations. However, in Chinese greenhouses, the limited operational space makes it difficult for large field machinery to be deployed, and there is a lack of specialized small-scale equipment. Many high-tech sprayers are still in the experimental stage and have not been widely adopted. Currently, chemical pest control in southern Xinjiang's greenhouse facilities mostly relies on devices like push-cart or backpack sprayers, which are affordable, easy to use, and have low maintenance costs. However, this traditional method of rain-like spraying is labor-intensive, covers a relatively small area of the plants, has poor adhesion to the undersides of leaves, leads to pesticide waste, and can cause pesticide pollution in greenhouse soil, posing risks to human health. As a result, the pest control effects of existing spraying machinery are unsatisfactory, affecting tomato growth and yield, and impeding the development of the greenhouse tomato industry in southern Xinjiang.

To address the issues of low mechanization, high labor intensity, short spraying range, poor atomization quality, and insufficient adhesion in pest control operations in southern Xinjiang's greenhouse tomato facilities, this study aims to integrate advanced electrostatic spraying technology into existing plant protection machinery, effectively combining agricultural machinery and agronomy. We will develop an electrostatic sprayer suitable for greenhouse tomatoes, design the core components, and conduct performance testing on the prototype. This will enhance the effectiveness of greenhouse pest control operations, improve pesticide utilization efficiency, promote the mechanization of greenhouse plant protection, and provide technical support for the advancement of electrostatic spraying in southern Xinjiang's greenhouse tomato industry.

9.2 Overall Design of the Sprayer

Based on field research conducted in greenhouse facilities in southern Xinjiang and a thorough review of relevant domestic and international literature on greenhouse spraying machinery, we gained a clear understanding of the farmers' requirements and actual needs for spraying equipment. To achieve effective plant protection management within limited operational space, ensure efficient droplet penetration and adhesion, and reduce environmental pollution, this section proposes the overall design of an electrostatic spraying machine powered by electric drive for greenhouses. The mechanical structure of the sprayer is also determined.

9.2.1 Design Requirements

Currently, greenhouse cultivation practices are hindered by outdated production concepts and low management levels, leading to weak plant growth and frequent outbreaks of major diseases, resulting in significant economic losses. Therefore, considering the greenhouse environment for tomato cultivation in Aksu, Xinjiang, this study focuses on the design and research of a sprayer tailored to tomato cultivation on trellises.

The electric-powered electrostatic sprayer for greenhouses should meet the following requirements, with the specific design as follows:

(1) Structural Requirements: The overall size of the sprayer must be suitable for operation in the narrow spaces of greenhouses, preventing damage to the crops and other structures. The proposed chassis dimensions for the sprayer are 1000 mm in length, 550 mm in width, and 600 mm in height, ensuring good maneuverability.
(2) Spraying Requirements: The sprayer should be adaptable to different growth stages of the plants, with flexible nozzle installation and adjustable angles to meet varying operational height requirements.
(3) Stability Requirements: When loaded with pesticides, the electrostatic sprayer should be able to move smoothly along predetermined paths during plant protection operations without damaging greenhouse structures or crops.
(4) Application of Electrostatic Spraying Technology: By incorporating electrostatic spraying technology, the sprayer aims to improve the deposition rate of spray droplets in the greenhouse, reduce pesticide usage, minimize soil pollution, and increase labor efficiency.

The main performance indicators are as follows:

(1) The operating speed of the sprayer is 0.5 m/s.
(2) The sprayer's load capacity meets the pesticide requirements for greenhouse tomatoes.
(3) The sprayer is capable of changing motor direction, adjusting flow rate, and modifying operational height.

9.2.2 Whole Machine Program

In the overall design of the sprayer, it can be divided into three main components: the spraying device, the power source, and the transmission and connection system. Based on data review and field research, the design and analysis were conducted with the goal of optimizing the sprayer's performance.

Spraying device: Hydraulic spraying can be used for both long-distance and short-distance applications. It is convenient to operate and ensures even pesticide application, but it also suffers from pesticide wastage. On the other hand, electrostatic spraying ensures uniform droplet distribution, enhances pesticide coverage, and improves pesticide utilization efficiency. To mitigate the issue of nozzle adsorption caused by electrostatic effects, a combination of air-assisted and electrostatic spraying is proposed.

Power source: Internal combustion engines provide strong power and high driving force by converting internal energy into mechanical energy. However, they are heavy and can cause environmental pollution during greenhouse operations. Electric motors, in contrast, are lightweight, environmentally friendly, easy to operate, and more suitable for small-scale plots.

Transmission and connection system: Given the limited operational space inside greenhouses, large trailer-mounted or semi-mounted sprayers are not suitable due to their size and difficulty in maneuvering. A self-propelled sprayer, which is agile and easy to operate, better meets the requirements for working in confined spaces.

To meet different operational needs, the structure of the electric-powered electrostatic sprayer for greenhouses includes two spraying modes: multi-nozzle vertical rod spraying and automatic hose-reel single-nozzle spraying. The main components consist of a remote-controlled wheel-rail chassis, electrostatic spraying device, electronic control system, and an automatic hose-reel system. The automatic hose-reel single-nozzle spraying mode is an auxiliary operation and will not be elaborated further in this paper.

The remote-controlled wheel-rail chassis allows the sprayer, loaded with pesticides, to move stably along the track under the control of the operator. The electrostatic spraying device is powered by a lithium battery, with a high-voltage electrostatic generator that charges the pesticide droplets, causing them to be influenced by the electric field force. This establishes an electrostatic field between the crops and the nozzles, creating an "electrostatic wrap" effect for efficient crop protection spraying. The main control system, powered by a battery, enables manual or automatic operation. Flow control for the sprayer is achieved through a valve connected to the electrostatic nozzles on the spray boom, while the walking mechanism is controlled by an electric motor.

9.3 Structural Design of the Whole Machine

Based on the overall design and requirements, the mechanical structure model of the sprayer is shown in Fig. 9.1. It mainly consists of a walking chassis, spraying device, air-assist unit, and automatic hose-reel system. The walking chassis includes a battery pack, sensors, drive mechanism, and auxiliary components. The electrostatic spraying device consists of a spray boom, nozzles, pump, controller, pesticide tank, electrostatic generator, and connectors. The air-assist unit includes a fan, power supply, control board, switch, and nozzle adjustment rod. The automatic hose-reel system comprises a controller, drive wheel, driven disc, and guide rod. To facilitate the sprayer's relocation and adapt to different site conditions, it employs a combination of regular rubber wheels and T-shaped guide wheels, with a rear-wheel drive system. The front features two rubber wheels and two lateral guide wheels. The sprayer moves along a laid track, where the drive wheels power the driven wheels for movement, and the guide wheels ensure straight-line travel. Sensors installed on the walking chassis allow the machine to automatically return when it reaches the end of a row.

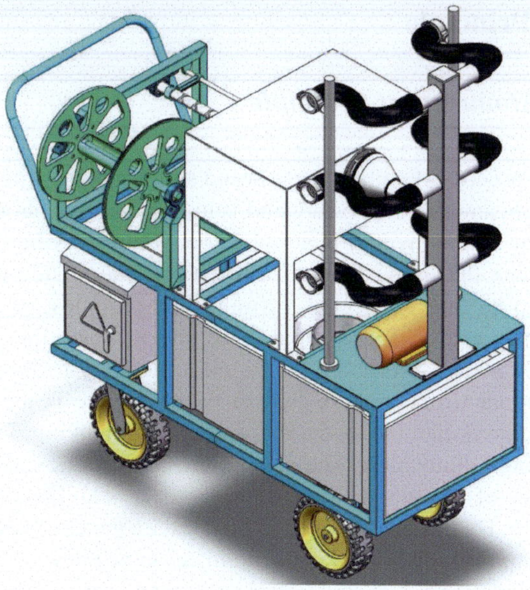

a. Axonometric drawing of an electrostatic sprayer

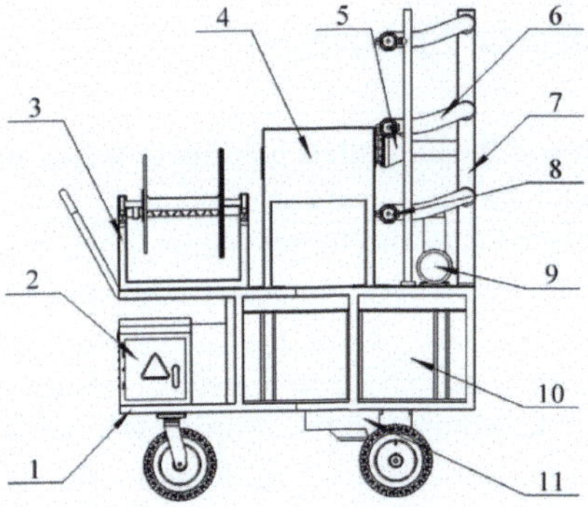

b. Main view of electrostatic sprayer

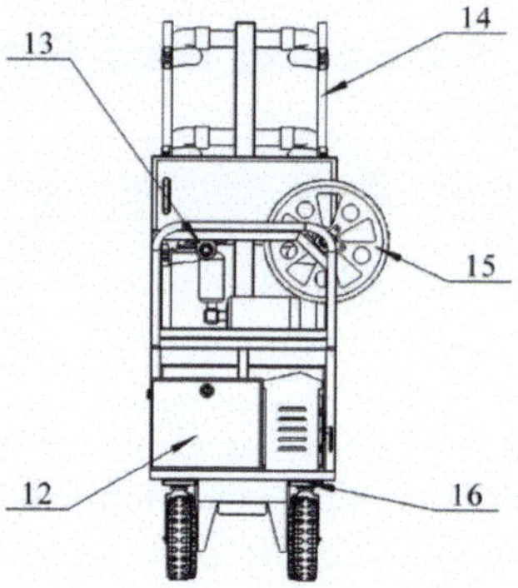

c. Side view of electrostatic sprayer

1.frame 2.control cabinet 3.hose reel movable frame 4.electrostatic generator cabinet 5.wind conveyor 6.wind conveying pipe 7.wind conveying pipe support 8.electrostatic nozzle 9.water pump 10.Medicine box 11.Power motor compartment 12.Battery cabinet 13.Hose reel reciprocating mechanism 14.Electrostatic nozzle support frame 15.Hose reel 16.Universal wheel support seat

Fig. 9.1 Schematic structure of facility tomato electrostatic sprayer

9.4 Working Principle

9.4.1 Liquid Atomization Principle

Atomization is the process of converting a liquid substance into tiny airborne particles through a certain amount of pressure and velocity. The greenhouse sprayer used in southern Xinjiang adopts hydraulic atomization, where a hydraulic pump generates pressure, causing the liquid to enter a turbulent state inside the nozzle. This allows the liquid to be rapidly ejected, forming fine particles, or mist. Hydraulic atomization primarily uses the centrifugal force generated by the liquid under the pump's action to break and accelerate the liquid, ultimately forming tiny droplets. This process is similar to the ripples created when a stone is thrown into water, where the liquid behaves like the stone, deforming and breaking due to centrifugal force, thus achieving atomization. The size of the droplets depends on the thickness and uniformity of the liquid layer, as well as the stability of the liquid and the breaking process. Hydraulic atomization nozzles are generally divided into solid cone atomization nozzles and hollow cone atomization nozzles.

9.4.2 Principle of Wind-Fed Electrostatic Spraying

Traditional pesticide spraying methods have low adhesion rates, leading to pesticide wastage and potentially causing environmental pollution and ecological damage. When only a small portion of the pesticide adheres to the target plants, pests and diseases may not be effectively controlled, which could even enhance pest resistance and exacerbate the vicious cycle of plant diseases and pests. In greenhouse plant protection operations, when the sprayer pumps the pesticide from the tank and atomizes it through the nozzle, some of the pesticide adheres to the crop leaves, while a portion evaporates into the air or seeps into the soil, causing waste and reducing pesticide efficiency. The tomato sprayer in southern Xinjiang greenhouses adopts the air-assisted electrostatic spraying method.

The electrostatic effect is achieved by powering the electrostatic generator, creating an electrostatic field between the nozzle and the target. When the air-assisted device blows the pesticide through the charged air, the surface of the crops accumulates charges, forming static electricity. Electrostatic forces are conductive and can propagate through air molecules. As the air-assisted device blows the charged air toward the crops, the charges are transferred to the crop surfaces, creating an electrostatic field. This field can attract small particles in the air, such as pesticide droplets. Due to this electrostatic effect, the pesticide particles are attracted to the crop surface, enhancing pesticide adhesion. Additionally, electrostatic forces allow the pesticide particles to distribute more evenly across the crop surface, improving pest control efficiency. Electrostatic spraying also reduces pesticide drift and loss, thus minimizing environmental pollution (Fig. 9.2).

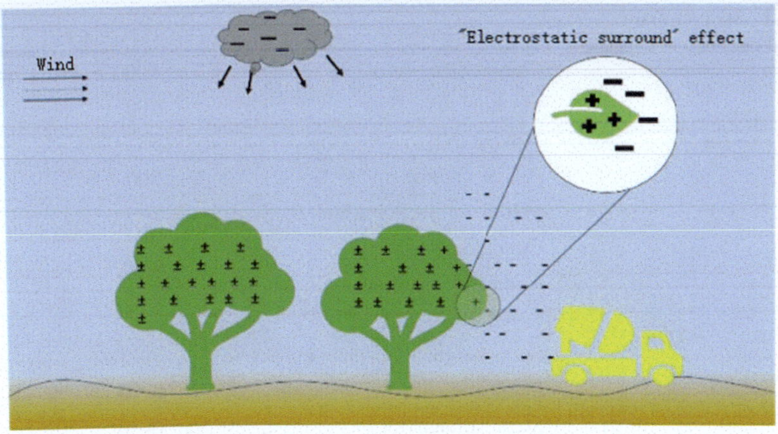

Fig. 9.2 Principle of electrostatic spraying

In practical applications, it was found that while electrostatic technology can charge the droplets, if the charged droplets do not reach the target in time, their charge may diminish. Therefore, this sprayer uses air assistance as a supplement. The liquid is atomized by the hydraulic pump, and the electrostatic spray charges the droplets. The airflow generated by the fan gives the droplets an initial velocity, allowing them to quickly reach the target and ensure the charged droplets act in time. This also improves adhesion in the deeper parts of the plants, enhancing the effectiveness of greenhouse plant protection operations and reducing pesticide waste.

9.4.3 Main Working Parameters

Based on the greenhouse tomato planting model, growth characteristics, and design requirements in southern Xinjiang, the main parameters of this model are determined as follows:

(1) Electrostatic voltage

The charging voltage of the high-voltage electrostatic generator is crucial for electrostatic spraying. Excessive voltage can affect the electrostatic ring effect and cause corona discharge, while insufficient voltage may not charge the droplets quickly enough, reducing the charge of the droplets. The operating range of the electrostatic voltage, adjustable via the high-voltage electrostatic generator control, is set to 0–20 kV.

(2) Spraying pressure

Spraying pressure directly affects the atomization of the droplets. If the pressure is too low, the multiple nozzles may not atomize the liquid properly, resulting in

Table 9.1 Sprayer main technical parameters

Main components	Parameters
Chassis size (mm)	1000 × 550 × 600
Machine mass (kg)	200
Voltage (V)	48
Fan air volume ($m^3 h^{-1}$)	500
Volume of medicine tank (L)	100
Operating height (m)	0.4–2.1
Number of combination nozzles	6
Speed ($m s^{-1}$)	0.5
Spray height adjustment range (cm)	60–180
Spray angle (°)	0–60
Static voltage (kV)	0–30
Spray pressure (MPa)	0.2–1.5
Nozzle type	Electrostatic nozzle

insufficient liquid atomization or large droplet sizes, leading to uneven coverage. If the pressure is too high, the droplet size may become too small, causing drift in the greenhouse environment. Therefore, the spraying pressure is set to a range of 0.2–1.5 MPa.

(3) Spray boom adjustment

The height of greenhouse tomato plants varies during different growth stages. In manual mode, the automatic hose-reel system has minimal impact on manual spraying. In the sprayer's automatic mode, the spray boom's nozzle adjustment system allows the nozzles to be positioned at different heights, meeting the automatic operation requirements for tomatoes at various growth stages. The adjustment range of the spray boom is set between 0.6 and 2.1 m, with the spray angle set between 0.60°. The main working parameters of the greenhouse tomato sprayer are shown in Table 9.1.

9.5 Design of Key Components of Electrostatic Sprayer

9.5.1 Electric Mobile Chassis Design

The designed sprayer is intended for use in standardized greenhouses, where the operational paths are relatively flat and the working conditions are uniform. Therefore, it is designed to move using an electric wheel system. Based on the working parameters of the sprayer, the structural dimensions of the electric mobile chassis are set as 1000 mm in length, 550 mm in width, and 600 mm in height. The sprayer

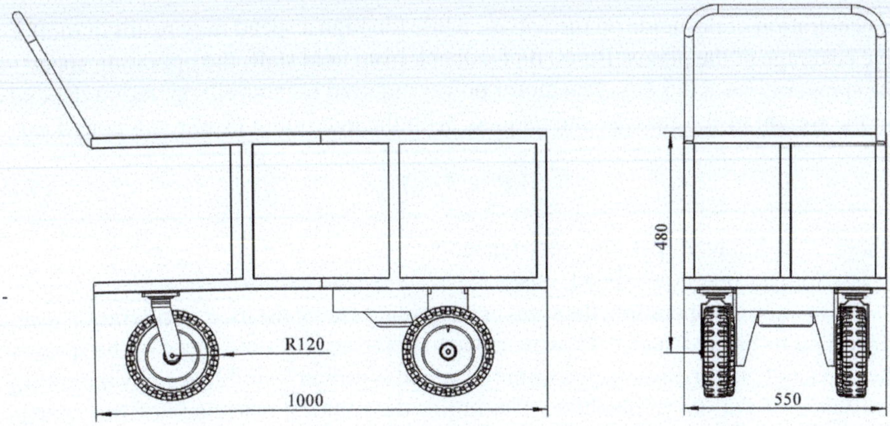

Fig. 9.3 Schematic diagram of the structure of the motorized mobile ground

is powered by a battery pack and driven by two-wheel motors, making it convenient to manufacture.

9.5.1.1 Chassis Frame Design

The sprayer's chassis must bear the weight of various components and withstand external forces during movement. Therefore, the design of the chassis must take into account the requirements for strength and rigidity. Additionally, considering the placement of the pesticide tank and the installation of other components, after a comprehensive analysis of the advantages and disadvantages of central-beam, composite, and side-beam chassis, the side-beam chassis, which provides larger carrying space and is easier to manufacture, is deemed most suitable. The material selected for manufacturing is Q235 ordinary carbon structural steel.

The 3D model of the chassis, complete with specific dimensions, was created using SolidWorks, as shown in Fig. 9.3. The chassis consists of two parts: the upper section is used to install the electrostatic spray unit, vertical spray boom, hydraulic pump, and other equipment, with a 200 × 150 mm overlapping rectangular plate reinforcement in the front. The lower section is used for the installation of the pesticide tank, battery, and control cabinet.

9.5.1.2 Chassis Drive Design

Common drive systems include hydraulic and electric motors. Considering the enclosed environment of the greenhouse, the sprayer utilizes a DC motor drive. If the sprayer operates too slowly during pest control operations, it may lead to excessive concentration of pesticide on the leaves, while too high a speed may result in ineffective coverage of the plants. Based on feedback from field staff, the operating speed of the sprayer is set at 0.5 m/s. The motor power required for smooth operation depends on the weight of the sprayer and its load, with the maximum power calculated as:

$$P = F_f \cdot V \tag{9.1}$$

where, F_f—Sprayer operating resistance, N;

v—Sprayer operating speed, m/s.

During normal operation, the sprayer experiences air resistance, acceleration resistance, and rolling resistance. However, due to the relatively slow speed of the sprayer, the impact of other resistances is minimal, and it is primarily influenced by the rolling resistance of the entire machine. Therefore, the running resistance of the sprayer during smooth operation is:

$$F_f = (M_C + M_Z)g \cdot \mu \tag{9.2}$$

where, M_c—Weight kg;

M_z—Sprayer load, g;

f—Friction coefficient, 0.25.

The self-weight of the chassis of the electrostatic sprayer is 60 kg, and the other loaded weight is about 170 kg, which is obtained by substituting the relevant parameters into the formula:

$$Ff = (60 + 170) \times 9.8 \times 0.25 = 563.5\,\text{N} \tag{9.3}$$

$$P = 563.5 \cdot 1 = 563.5\,\text{W} \tag{9.4}$$

Based on the calculations above and considering the power consumption of the motor, a 24 V 600 W DC variable-speed motor is selected. This motor is compact, reliable, and meets the operational requirements of the electrostatic sprayer.

9.5.1.3 Mobile Chassis Simulation Analysis

The chassis of the sprayer is prone to bending and deformation during spraying operations, which can affect its service life. Therefore, to ensure that the sprayer can operate normally and complete its spraying tasks, the strength and stiffness of the chassis are verified using ANSYS Workbench software before production. The specific analysis steps are as follows:

1. Use the Design Modeler module in ANSYS Workbench to create a simplified model of the sprayer chassis, omitting structural components that have minimal impact on finite element analysis.

2. Set the chassis material to Q235 ordinary carbon structural steel and perform mesh generation and solution for the model.
3. Open the Static Structural-Mechanical module in ANSYS Workbench and apply loads to the chassis model, including the weights of the battery, spray unit, pesticide tank, and hydraulic pump, while applying constraints to the wheels and chassis. As shown in Fig. 9.4a, the weight of the pesticide tank (A) is applied at the center of the beam, the weights of the spray unit and hydraulic pump (C) are applied to the front beam, and the weight of the battery (B) is applied to the support plate.
4. Calculate and analyze the stress and strain of the chassis under maximum load conditions, as shown in Fig. 9.4b, c.

From the stress distribution contour map of the chassis, it is observed that the maximum stress value of 16.112 MPa occurs at the contact point between the wheel frame and the support plate. This stress concentration is primarily due to the weight of the mounted equipment, indicating that the pesticide tank and spray boom significantly impact the internal stress of the chassis when fully loaded and stationary.

The displacement contour map under maximum load shows that the maximum displacement under static load is 0.18 mm. The greatest deformation displacement occurs at the contact area between the spray unit and hydraulic pump and the support plate of the chassis. Although this area lacks structural reinforcement, the displacement change is minimal. Sprayer chassis material is Q235, when the permissible safety factor selected as 1.5, then the permissible stress $[\sigma] = \sigma_s/\zeta = 156.67$ MPa, permissible deflection $[\omega] = L/500 = 1.1$ mm, L is the span of the lateral chassis of the sprayer, and the comparison shows that the mobile chassis of the sprayer meets the strength and stiffness design requirements.

9.5.2 *Design of Wind Conveying Unit*

The wind-assisted electrostatic sprayer in the southern Xinjiang facilities includes a wind machine, air duct, and other components. The air duct is made of flexible corrugated tubing, installed at the air intake of the wind-assisted spray unit, with a spray nozzle at the outlet. During operation, the fan directs airflow through the air duct to the corrugated tubing, and the mist is ejected from the electrostatic nozzles through the left and right outlets. The wind then blows the mist onto the leaf surfaces of the vegetable crops on both sides of the greenhouse rows.

When the wind force is too low, it cannot effectively flip the vegetable leaves, reducing coverage on both sides of the leaves, and the mist droplets are not fine enough to meet pest control requirements. Conversely, when the wind force is too high, the mist droplets, which are evenly dispersed due to electrostatic effects, can drift away and fail to reach the crop surfaces.

a. Load design diagram

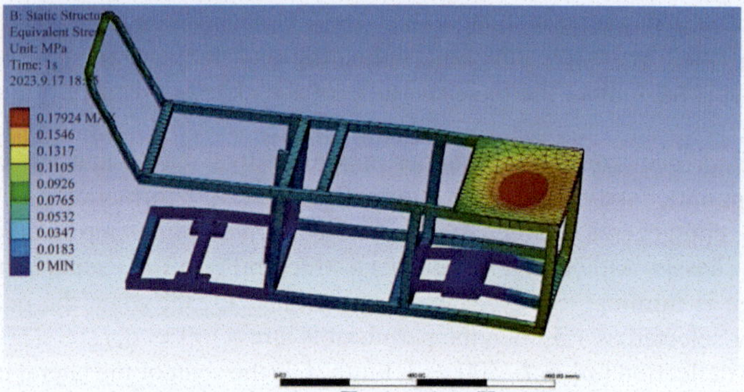

b. Stress cloud diagram

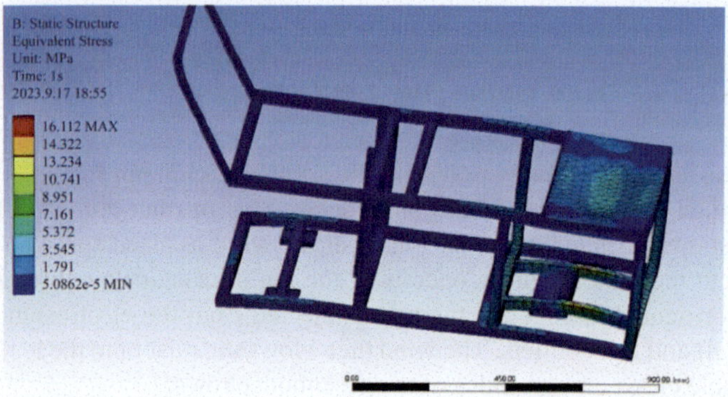

c. Strain cloud diagram

Fig. 9.4 Maximum load stress and deformation of motorized moving ground

9.5.2.1 Fan Parameterization

In greenhouse pest control operations, the main parameters for determining the fan are air volume and air pressure. Existing literature suggests that the air volume is determined by the replacement principle. For the wind-assisted electrostatic sprayer in the southern Xinjiang facilities, the replacement principle is that the charged mist ejected by the sprayer should replace all the air in the space around the vegetable leaf surfaces on both sides of the spray nozzles. The specific calculation formula is as follows:

$$Q \geq Vh/\psi \tag{9.5}$$

where:

Q is the air volume of the spraying machine, m^3/s;
v is the operating speed of the machine, m/s;
h is the height of vegetable plants, m;
l is the width of vegetable plants from the machine, 10.0–25.0 cm;
ψ is the air loss coefficient, 1.2–1.5.

The greenhouse environment has good airtightness with minimal external interference, so ψ is taken as 1.2. Based on the surveyed height of vegetable plants in the greenhouse (h) of 1.5 m and l of 0.15 m, and the average working speed of the sprayer being 0.5 m/s, the calculated air volume Q is 0.135 m^3/s, which is equivalent to 486 m^3/h.

When the fan sends air through the air duct, the pressure loss of the wind pressure mainly has dynamic pressure loss and static pressure loss, dynamic pressure loss calculation formula is as follows:

$$P = \rho v^2/2 \tag{9.6}$$

Static pressure loss contains pressure loss and friction loss, due to the duct is shorter resistance is smaller, so no longer consider, only calculate the pressure loss, the formula is as follows:

$$P = \lambda \rho v^2/2 \tag{9.7}$$

where:

P is the dynamic pressure loss, Pa;
P is the static pressure loss;
ρ is the air density, kg/m^3, taken as 1 kg/m^3;
v is the air velocity, m/s.
λ is the local pressure coefficient, and 1.3 is selected for this machine.

Using a handheld anemometer, the average wind speed at each outlet of the sprayer was measured to be 8 m/s. The total pressure loss of the fan was calculated to be 73.6 Pa. Based on the above analysis and discussions with suppliers, the final choice is

Table 9.2 Fan performance parameters

Name	Model	Voltage (V)	Power (W)	Air volume (m^3/h)	Pressure (Pa)	Speed (r/min)	Temperature (°C)
Centrifugal fan	SC3G190-DAN-72	12	32	500	350	2500	−25 to +60

the Sinen SC3G190-DAN-72 centrifugal fan, with specific performance parameters as listed in the Table 9.2.

9.5.2.2 Determination of Duct Parameters

In order to satisfy the spraying needs and cultivation patterns of vegetables in different growing periods, a total of six spray nozzles were designed at the left and right ends of the air ducts, with a duct height of 1.2 m and a coverage area of all air outlets:

$$S_z = Q/v \tag{9.8}$$

where, S_z is the coverage of the windrows, m^2;

Q is the air volume of the spraying machine, 0.135 m^3/s;

v is the outlet wind speed, 8 m/s, substituting into the formula to find the area of a single air outlet is 0.00028125 m^2, considering the structure of the electric ring is circular, so the radius of the outlet parameter is 30 mm.

9.5.3 Spray Unit Design

The spray unit ensures the supply and adjustment of pesticides during operations. The hydraulic pump pressurizes the liquid from the tank and supplies it to the nozzles through the pipelines on the spray boom. The spray unit mainly consists of two parts: the spray boom and the liquid supply system. The liquid supply system includes the hydraulic pump, pesticide tank, and water pipes.

9.5.3.1 Spray Stand Design

The design parameters of the spray support directly impact the effectiveness of greenhouse spraying. Based on field research and literature review, commonly used support structures include trapezoidal, portal, and rectangular shapes. Considering the requirements for pesticide application on vegetable crops at different growth stages and the narrow operational space in the greenhouse in southern Xinjiang, a rectangular spray support was chosen. By fixing bolts at different positions on the bottom of the spray support, the horizontal distance of the spray boom from the crops can be adjusted.

In the facility research in southern Xinjiang, it was found that the row spacing of greenhouse crops ranges between 600 and 700 mm. Based on the current chassis design width, the support width was chosen to be 500 mm. Additionally, to accommodate the need for spray angle adjustment, an adjustable spray head clamping device was designed, allowing the spray head to be adjusted within a range of 0 to 180°, effectively enhancing the coverage of the sprayer. According to the nozzle selection mentioned earlier, the spray angle of the nozzle is $\partial = 110°$. During pest control operations, when the sprayer moves between rows, the distance to the plants is approximately $L = 250$ mm. Ignoring other factors affecting the spraying operation, the coverage area S of the spray head can be calculated, where $0 < P < S$:

$$S = 2L\tan(\partial/2) \tag{9.9}$$

The number of nozzles installed on the vertical spray bar N is:

$$N = H/P \tag{9.10}$$

Among them, H is the height of the vertical spray bar; nozzle spacing $P = S/2$, the calculation of the number of nozzles required to take 3, nozzle spacing 390mm, as shown in Fig. 9.5.

9.5.3.2 Liquid Supply System Design

(1) Hydraulic pump

According to the "Mechanical Design" handbook, when selecting a hydraulic pump, both the volume and the requirements for displacement and pressure must be considered. The hydraulic pumps currently used in sprayers include diaphragm pumps, piston pumps, and gear pumps. Based on the actual working needs of greenhouse spraying, the required spray flow rate for the sprayer can be calculated using the following formula:

$$Q = \frac{MV_F B}{600} \tag{9.11}$$

where: Q—Total flow for all nozzles, L/min;

M—Amount of liquid used, L/ha;

V_F—Travel speed of the sprayer during operation, km/h;

B—relief shot, m.

Check the relevant literature can be seen, greenhouse spraying machinery spraying volume for $M = 200$ L/ha, electrostatic spraying machine traveling speed for $V_F = 1.8$ km/h, according to the greenhouse double spraying the actual, the spray width of

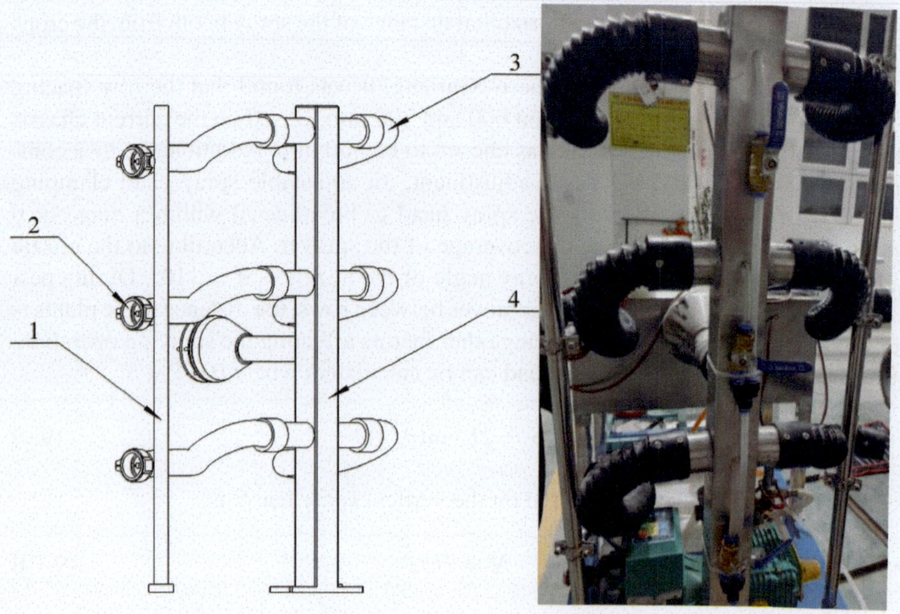

1.Electrostatic nozzle holder 2.Electrostatic nozzle 3.Air duct 4.Air duct holder

Fig. 9.5 Spray bracket structure and object

B to take 3.4 m, will be related to the parameter into the formula (9.11) calculations can be obtained: $Q = 2.04$ L/min.

Selected hydraulic pump parameters, as shown in Table 9.3.

(2) Medicine cabinets

The pesticide tank is an essential component of the tomato sprayer, responsible for storing and supplying the pesticide during operation. It is the largest part of the sprayer in terms of volume. Since the sprayer needs to navigate between tomato rows, its overall size is limited. Frequent changes of pesticide increase labor intensity, so the size of the tank directly affects the machine's workload and efficiency. A well-designed tank can minimize the number of refills, reduce labor intensity, and improve

Table 9.3 Hydraulic pump parameters

Parameters	Indicators
Model	P16-A0-F-R
Working pressure (MPa)	0–4
Displacement (L/min)	4–16
Dimension (mm)	214 × 130 × 142
Speed (r/min)	500
Weight (kg)	6

efficiency. To prevent the tank from falling off due to inertia during movement, the tank is designed in a long-strip shape, allowing it to be secured in place by slots on the frame, thus increasing stability and reliability. As many pesticides are corrosive, the tank is made of polyethylene, which is resistant to corrosion, moisture, and aging.

According to the "Agricultural Machinery Design Handbook," the liquid volume for the electrostatic sprayer is calculated as follows: For a tomato facility with six nozzles on each side, at a working pressure of 0.6 MPa, the flow rate for each nozzle qq is 0.4 L/min. The total flow rate for all nozzles is given by:

$$Q_y = q \cdot N = 0.4 \times 6 = 2.4\,\text{L/min} \tag{9.12}$$

According to the previous design, the working speed v of the spraying machine is 0.5 m/s. The greenhouse has 100 rows of tomatoes, with each row approximately 12 m long. Let S be the distance traveled by the spraying machine, then the volume of the spray tank V_Y can be calculated as follows:

$$V_Y = \frac{Q_Y S}{60\text{v}} = 96\,\text{L} \tag{9.13}$$

Therefore, in the design, a 100 L spray tank was selected for the machine, which is sufficient to complete pest control operations in a single greenhouse. This choice reduces the frequency of refilling the pesticide and enhances the efficiency of pest control.

9.5.4 *Electrostatic Device Design*

9.5.4.1 Charging Method of Electrostatic Atomization

The safety and reliability of droplet charging are crucial in the design of electrostatic spraying systems. The quality of the charge directly affects the effectiveness of electrostatic forces during pest control operations. Currently, the most common charging methods include corona charging, triboelectric charging, direct conductive charging, and inductive charging. However, triboelectric charging is rarely applied in pesticide electrostatic spraying. The equivalent circuit principles of the remaining three charging methods are shown in Fig. 9.6.

The charging voltage for contact charging typically ranges between 20 and 30 kV. In this method, the pesticide or nozzle is directly connected to one end of a high-voltage power source for charging, while the other end is grounded. This method achieves good charging results, but it places high demands on the insulation of the electrostatic spraying system. Since the entire spraying system is electrified, there are certain safety risks. The equivalent circuit principle for this charging method is shown in Fig. 9.6a. Corona charging operates at voltages above 30 kV. It utilizes discharge

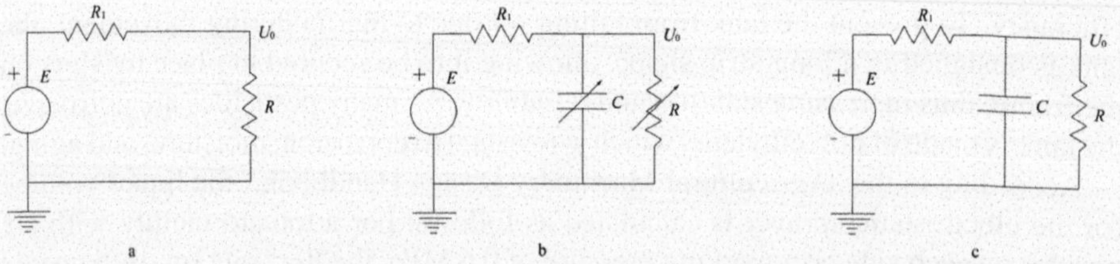

Fig. 9.6 Equivalent circuit schematic of electrostatic atomization charging method

from the tip of a high-voltage electrode to ionize the surrounding air, making it relatively easy to operate. This method is also applicable to standard nozzles. However, due to the high voltage, it carries a certain level of risk. The equivalent circuit principle for corona charging is shown in Fig. 9.6b. Inductive charging typically operates within a voltage range of 1–15 kV. The high-voltage power source is placed between the electrode and the nozzle, forming an electrostatic field near the electrode. Insulation is only required at the induction electrode, making this method safer. The equivalent circuit principle for inductive charging is shown in Fig. 9.6c. Considering factors such as the spraying machine's energy consumption and safety requirements, the electrostatic device was designed using inductive charging, with the charging principle illustrated in Fig. 9.7. The electrostatic spraying system is powered by a 24 V lithium battery. Since the high-voltage electrostatic generator operates at 220 V, a step-up converter is needed to convert the low-voltage DC to 220 V AC. To ensure the safety and reliability of the high-voltage electrostatic generator, each nozzle is connected in parallel to the high-voltage generator, with wires connecting the electrode's small aperture to the electrostatic generator. A physical diagram of the electrostatic device is shown in Fig. 9.8.

9.5.4.2 Electrostatic Printhead Design

The nozzle used in the spraying machine needs to simultaneously fulfill both electrostatic and air-assisted functions, requiring modifications to the original nozzle design. Currently, a variety of nozzles are used, with common types including atomizing nozzles and centrifugal nozzles. Generally, these can be classified into two main categories: round nozzles and fan-shaped nozzles. Fan-shaped nozzles have adjustable angles, concentrated flow, and higher central pressure, with the pressure gradually decreasing as it diffuses outward. They are typically used for cleaning tasks. Round nozzles, on the other hand, produce cone-shaped droplets that are more evenly distributed.

Considering the advantages and disadvantages of each type, a rotating atomizing nozzle was ultimately selected for the design, as shown in the diagram of the unmodified atomizing nozzle. Based on the principle of inductive charging, a circular electric ring was designed, insulated using engineering plastic. The external surface

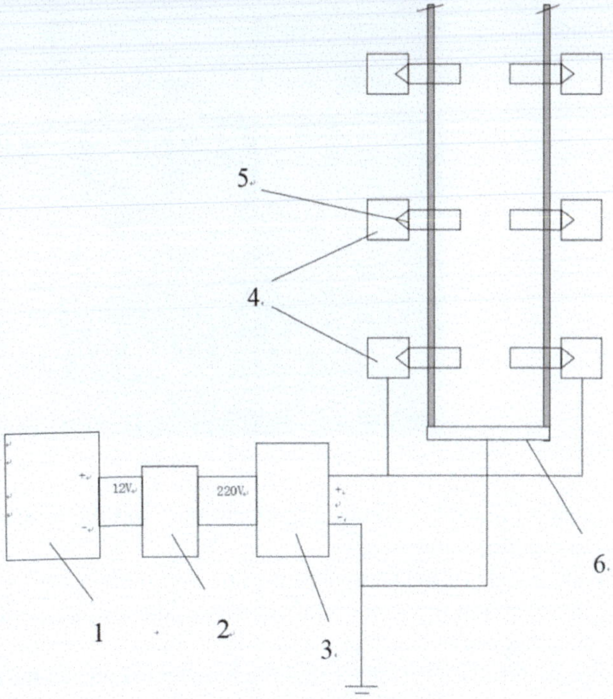

1.Battery 2.Converter 3.High-voltage electrostatic generator 4.Circular electrode 5.Nozzle 6.Spray bar

Fig. 9.7 Fog droplet induction charging working principle schematic

Fig. 9.8 Physical drawing of electrostatic device

Fig. 9.9 Electrostatic spray nozzle physical picture

is equipped with small holes for connecting electrodes, and the nozzle is mounted on the air-assisted pipeline using screws.

The electrostatic nozzle of the spraying machine is equipped with an air-assisted airflow component, which imparts an initial velocity to the droplets, enhancing their adhesion under the influence of the electrostatic force. Holes were made in the electric ring and insulation material to fix screws, making it easier to connect the high-voltage terminal to the electric ring. The physical structure and schematic diagrams of the air-assisted electrostatic nozzle are shown in Figs. 9.9 and 9.10.

The structural dimensions of the electrostatic nozzle electrode, as well as the spacing between the nozzle and the electric ring, significantly affect the distribution of the spraying electric field during the design process. As shown in Fig. 9.11, the nozzle-electric ring model illustrates the key parameters of the electrostatic nozzle. Based on the design requirements, the table lists the proposed parameters for the electrostatic nozzle (Table 9.4).

9.5.4.3 Electrostatic Generator Design

The role of the electrostatic generator is to produce high-voltage static electricity. Based on the working voltage range for inductive charging and literature reviews, a 10 kV voltage is sufficient to meet the pest control needs of a greenhouse electrostatic sprayer. Commonly used electrostatic generators have an input voltage of 220 V, but they can be modified to be powered by the sprayer's built-in 12 V or 24 V battery. This sprayer is equipped with the JD-SFQ092 high-voltage electrostatic generator, which operates at 220 V and has an output voltage range of 0–30 kV. The physical electrostatic generator is shown in Fig. 9.12.

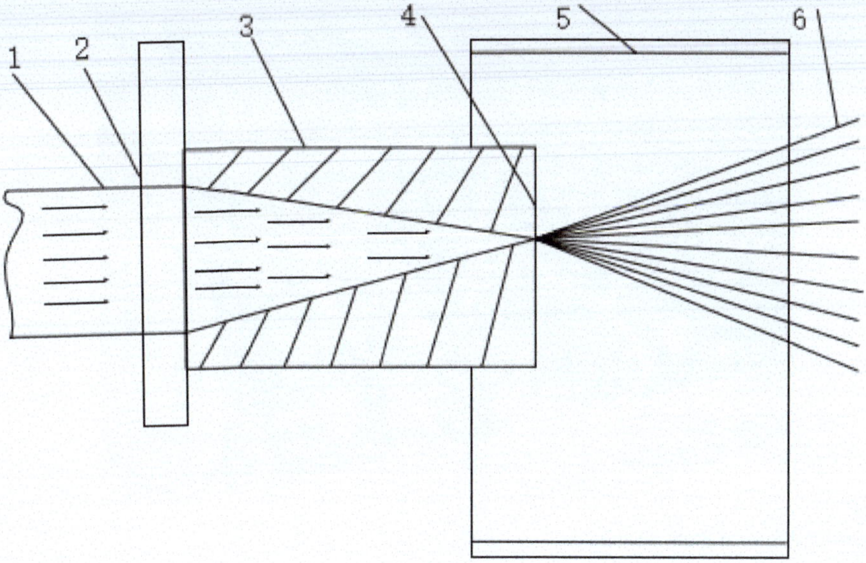

1.Tube 2.ventilation assembly 3.nozzle body 4.nozzle 5.ring electrode 6.charged droplet

Fig. 9.10 Working diagram of electrostatic printhead

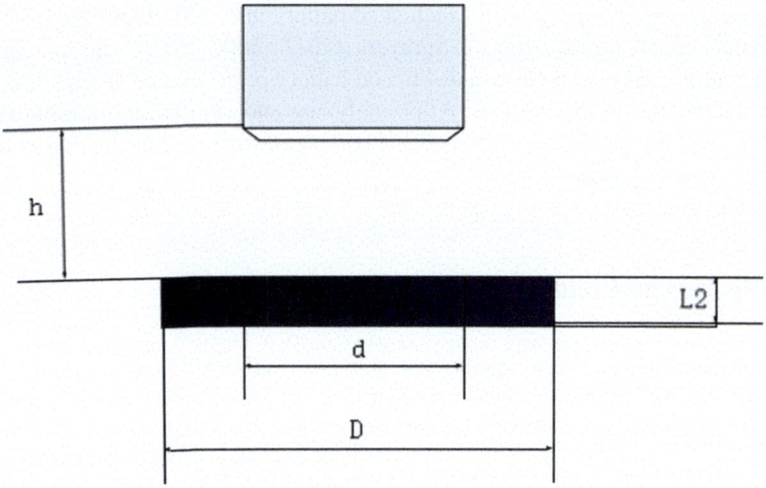

Fig. 9.11 Electrode-nozzle model

Table 9.4 Electrode-nozzle model parameters

Parameters	D (mm)	d (mm)	h (mm)	Angle	L2 (mm)
Value	42	32	5	110°	10

Fig. 9.12 Physical Static Generator

9.5.4.4 Power Supply Selection

The power supply system for the electrostatic spraying device is separate from the chassis motor power supply, with a dedicated battery installed. According to "Technical Requirements for Electrostatic Sprayers (GB/T 33006-2016)," the sprayer must operate continuously for no less than 4 h, and battery performance degradation must remain within the specified range. After analyzing and comparing multiple brands of power supplies, the Camel 12 V 100 AH battery was selected as the power source for the electrostatic generator.

9.6 Prototype Field Test and Analysis

In greenhouse pest control operations, achieving effective results requires that droplets cover both the upper and lower surfaces of plant leaves. This ensures that pests on the leaves are controlled, preventing them from damaging crops. Additionally, the droplet size must be appropriate to ensure pest control effectiveness. According to agricultural experts, different droplet sizes have varying levels of lethality against pests. In greenhouses, leaf-surface crawling pests are the primary concern affecting plant growth. The optimal droplet size for effectively killing crawling pest larvae on leaves ranges from 30 to 150 μm. If the droplet size is too small, it tends to remain suspended in the air, preventing it from reaching the leaf surface in time. Conversely, if the droplet size is too large, it may be shed from the leaf surface or fall to the ground due to gravitational forces, resulting in the loss of the solution before it can be absorbed. Therefore, spray deposition and droplet size are critical indicators for evaluating the performance of a sprayer in pest control operations.

To clarify the significance of various factors on the effectiveness of electrostatic spraying for pest control, the electrostatic sprayer was used as a test prototype to study the effects of key factors such as electrostatic voltage, spray pressure, and spraying distance on pest control performance. Through multi-factor analysis, the optimal parameters were determined, providing a basis for selecting efficient operating parameters for the electrostatic sprayer.

9.6.1 Test Material

9.6.1.1 Purpose of the Test

To ensure that the greenhouse electrostatic spraying prototype operates in an optimal state and to verify whether the design parameters of the prototype meet farmers' performance requirements for pest control operations, spray droplet coverage was selected as the evaluation criterion. Field experiments were conducted to determine the optimal operating parameters of the sprayer. These experiments aim to achieve effective pest control while simultaneously improving the reliability of the prototype during operation.

9.6.1.2 Test Condition

The experiment was conducted in a tomato greenhouse located in Aral City, First Division of the Xinjiang Production and Construction Corps. The conditions were calm, with a temperature of 14 °C and air humidity of 82%. The tomatoes were planted in double rows, and the experiment was carried out during the peak period of tomato growth, which coincided with the critical stage of pest and disease control. Standardized parameters such as row spacing, plant spacing, and the height of the main stems were measured and recorded. To enhance data reliability, multiple measurements were taken and averaged. The basic parameters of the tomato plants are shown in Table 9.5.

Table 9.5 Basic parameters of greenhouse tomato

Vegetable parameters	Row spacing (cm)	Plant spacing (cm)	Plant main stem height (cm)
1	71	32	162
2	69	40	187
3	70	36	172
Average value	70	36	174

Fig. 9.13 Pictures of the test site

9.6.1.3 Test Material

The experimental equipment included: an electrostatic sprayer, a measuring tape, water-sensitive paper, sample bags, water, a temperature and humidity meter, a notebook, plastic clips, plastic gloves, and other necessary tools. The nozzle spacing on the vertical spray rod of the sprayer was adjusted to 300 mm. The on-site greenhouse experiment setup is shown in Fig. 9.13.

9.6.2 Test Methods

9.6.2.1 Pilot Program

The experiment followed the technical standards outlined in "GB/T 33006-2016 Technical Requirements for Electrostatic Sprayers" and "DB32/T 3574-2019 Technical Specifications for the Evaluation of Electrostatic Sprayer Operation Quality." On-site testing was conducted with three replicates according to a randomized block design within the greenhouse. In the southern part of the greenhouse, an experimental area was delineated into three blocks. Each block consisted of five crop rows, with three of these rows used for the experiment. To prevent potential contamination, the rows were separated by a line, as shown in Fig. 9.14.

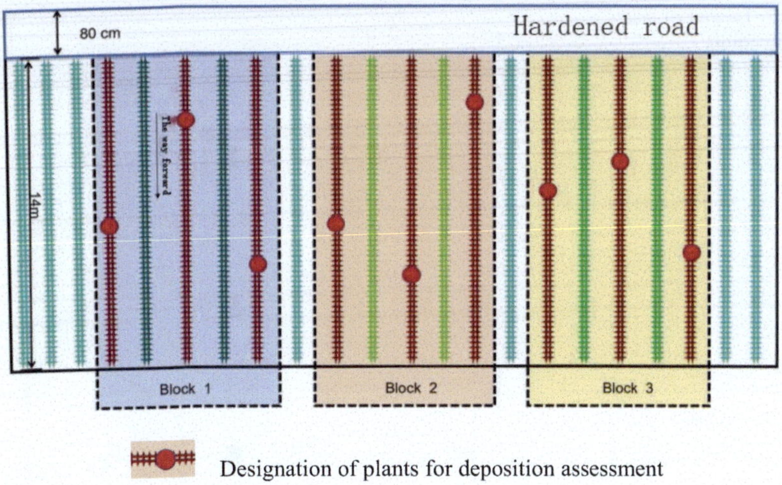

Fig. 9.14 Trial plots and sampling crop rows

In the standardized planting of greenhouse tomatoes, which exhibits certain symmetry, sampling for the experiment was conducted on only one side of the collection rack. The layout of the sampling points is shown in Fig. 9.15. Due to limited space in the inter-row working pathways, the vertical spray rod was fixed at a set distance from the crop axis. Sampling was conducted at three horizontal planes parallel to the ground for one side of the tomato plants, with vertical intervals of 20–50 cm, 80–110 cm, and 140–170 cm, respectively. At each sampling point, two pieces of water-sensitive paper were placed and secured with plastic pins on both sides of the leaves, resulting in a total of six sampling points. The sampling points were numbered from top to bottom as A1, A2, B1, B2, C1, and C2. After allowing the water-sensitive paper to stabilize, it was removed and sealed in sequence from nearest to farthest using plastic gloves. The average of the values from the upper, middle, and lower positions within the same plane was used as the result for each sampling point.

9.6.2.2 Test Indicators

The index of the electrostatic spray test is the percentage of droplet coverage on the front and back of the plant, specifically the ratio of the droplet area covered by the droplets on the specimen area during the spraying operation to the total area on the specimen used for counting, with the following equation

$$C = \frac{A_S}{A_P} \times 100\% \tag{9.14}$$

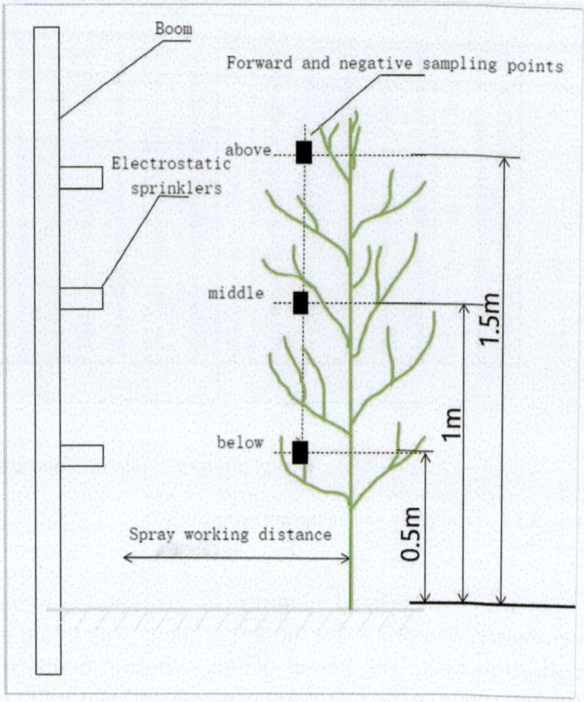

Fig. 9.15 Layout of test sampling points

where, C is spray droplet coverage, %;

A_S is the number of pixels in the specified area of the spray droplet;

A_P is the number of pixels to count the total area.

9.6.2.3 Measurement Methods

According to the experimental protocol, water-sensitive paper with dimensions of 30 mm × 60 mm was placed on the tomato plants. The spraying tests were conducted by varying experimental factors. After achieving stable spraying, the electrostatic spraying device was moved across the tomato plants. Once the water-sensitive paper on the plants had settled and dried, it was collected and scanned at 600 dpi in PNG format using a scanner.

For analysis, computer software was used to process the scanned images. The specific analysis process is illustrated in Fig. 9.16. To reduce the impact of edge contamination on the accuracy of the test results, only the central stable region of 20 mm × 20 mm on the water-sensitive paper was analyzed for droplet coverage, as shown in Fig. 9.17.

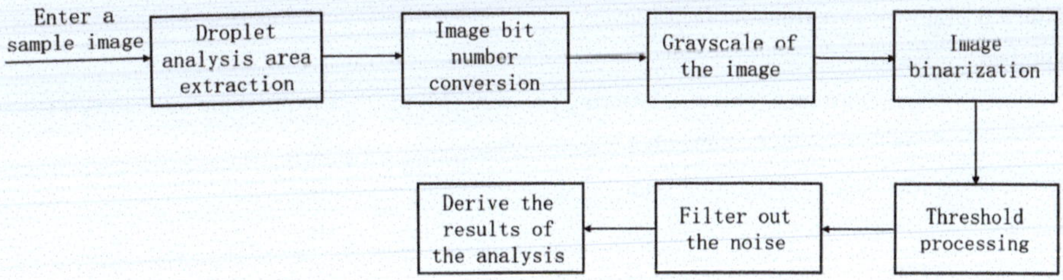

Fig. 9.16 Calculating droplet coverage process

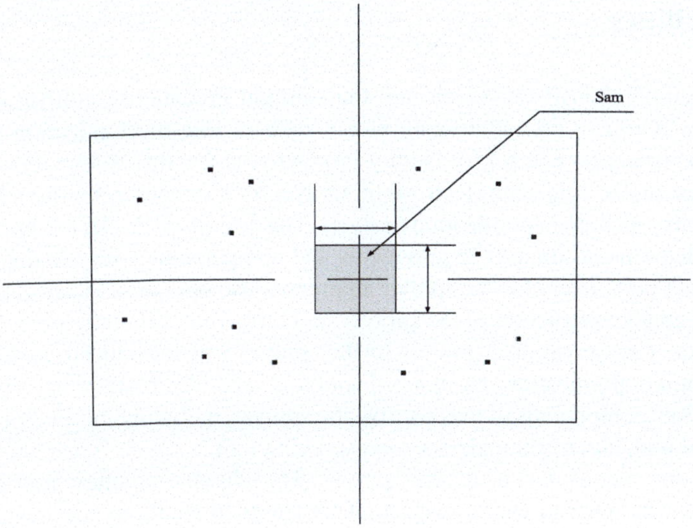

Fig. 9.17 Droplet adhesion analysis area

9.6.2.4 Test Factors and Levels

Based on the preliminary literature review and theoretical simulation analysis of electrostatic spraying, electrostatic voltage, spray distance, and spray pressure were selected as experimental factors. Initially, single-factor experiments were conducted, with other factors held constant to determine their individual effects. To optimize the electrostatic spraying parameters among multiple factors, a response surface methodology was employed to further refine the experiments and identify the optimal combination of parameters for spraying operations. The levels of the experimental factors are listed in the table. During the experiments, the electrostatic sprayer was operated at a speed of 0.5 m/s, and the fan power was kept constant. The average wind speed, measured using an anemometer, was 8 m/s (Table 9.6).

Table 9.6 Table of experimental factor levels

Number	Factor		
	Spray distance (cm)	Working pressure (MPa)	Electrostatic voltage (kV)
1	25	0.4	2
2	50	0.6	6
3	75	0.8	10
4	100	1.0	14

9.7 Research Findings on Smart Sprayers for Desert Gobi Facilities

Based on comprehensive research into the sunlight greenhouse cultivation model in southern Xinjiang and addressing issues such as low mechanization and high labor intensity in pest control for facility-grown tomatoes, this study focused on the development and testing of an electrostatic sprayer for tomatoes in southern Xinjiang. The study also included optimization analysis. The design of the electrostatic sprayer was completed based on user requirements and comparisons with existing models. The working parameters for the sprayer in greenhouse operations were established, and the mechanical structure of the sprayer was three-dimensionally modeled using SolidWorks. The operational process of the sprayer was introduced, and the principles of liquid atomization, air-assisted spraying, and electrostatic operation were analyzed. Key components of the electrostatic sprayer, including the electric chassis, air-assisted unit, electrostatic device, and spraying unit, were designed and selected based on the overall design of the sprayer. The chassis structure was designed according to the growth characteristics of tomatoes in southern Xinjiang, and the motion parameters of the chassis drive were calculated. A 3D model of the chassis was created using SolidWorks, and static analysis of the chassis load during operation was performed using ANSYS software, verifying that the chassis met strength and rigidity requirements. Considering the greenhouse working environment, calculations for the sprayer's air volume and pressure determined the fan performance parameters, leading to the selection of the appropriate fan. Calculations for the duct parameters were also completed. The width, length, and nozzle spacing of the sprayer were determined, and the theoretical flow rate during operation was calculated, leading to the design and selection of the hydraulic pump and tank volume. Based on the analysis of electrostatic charging methods, inductive charging was selected for the electrostatic system. The design and selection of the nozzle, electrostatic generator, and power supply were completed.

The intelligent spraying machine developed in this study provides a valuable solution for facility agriculture in desert and Gobi regions, enhancing spraying efficacy and offering technical support and reference for agricultural production in similar environments in the future.

Chapter 10
Research on the Mechanized Operation Mode of the Whole Process of Facility Agriculture in Desert and Gobi

Jiahui Xu, Jianfei Xing, Long Wang, Xufeng Wang, Yalong Song, and Jiacheng Xin

10.1 Introduction

As China's agricultural modernization progresses, mechanization becomes increasingly vital for enhancing production efficiency, reducing costs, and fostering sustainable development. Full-process mechanization in desert and gobi facility agriculture is designed to promote widespread mechanization in these regions, boosting production efficiency and quality while supporting the upgrading of the agricultural industry and regional economic growth. Desert and gobi areas, characterized by their vast and unique geography, face significant challenges such as extreme climates, water scarcity, and barren soil. Nevertheless, advancements in facility agriculture and technology reveal substantial potential in these regions. With scientific design and strategic planning, desert and gobi facility agriculture can achieve efficient resource

J. Xu · J. Xing · L. Wang · X. Wang (✉) · Y. Song
College of Mechanical and Electrical Engineering, Tarim University, Alar 843300, China
e-mail: wxfwyq@126.com

Modern Agricultural Engineering Key Laboratory at Universities of Education Department of Xinjiang Uygur Autonomous Region, Alar 843300, China

Xinjiang Production and Construction Corps (XPCC) Key Laboratory of Utilization and Equipment of Special Agricultural and Forestry Products in Southern Xinjiang, Alar 843300, China

L. Wang
e-mail: 120140002@taru.edu.cn

J. Xin
Key Laboratory of Smart Agriculture System Integration, Ministry of Education, Beijing 100083, China

Key Laboratory of Agricultural Information Acquisition Technology, Ministry of Agriculture and Rural Affairs, Beijing 100083, China

C. Hu et al. (eds.), *New Technologies Applied in Gobi Desert Facility Agriculture*, Smart Agriculture 11, https://doi.org/10.1007/978-981-96-1031-0_10

use and a stable food supply, contributing positively to the broader development of China's agriculture and regional economic enhancement.

Traditional agricultural methods in desert and gobi regions face significant challenges, including high labor intensity, low production efficiency, and inconsistent quality, which fall short of modern demands for efficiency, sustainability, and high-quality outputs. Therefore, promoting full-process mechanization in these areas is essential. This technology not only dramatically increases production efficiency and product quality but also effectively reduces production costs, minimizes resource waste, and supports sustainable agricultural development. Research into mechanization models for desert and gobi facility agriculture aims to provide scientific and standardized technical guidance. It will offer detailed insights into planting design, mechanization processes, technical requirements, equipment, operational standards, and key considerations to address challenges such as harsh environments, resource scarcity, poor machinery adaptability, high operational difficulty, and low profitability. This guidance will assist farmers and technicians in performing mechanized operations correctly and efficiently, thus enhancing mechanization levels in facility agriculture. The research should take into account the unique geographical and climatic conditions of these regions, leveraging advanced agricultural technologies to optimize design and layout for effective mechanization. For example, addressing water scarcity can involve adopting water-saving irrigation techniques and efficient water use technologies, complemented by mechanized equipment for precise irrigation to significantly improve water use efficiency. Additionally, utilizing drought-resistant and salt-tolerant crop varieties can enhance crop survival and yield in challenging environments.

Research on the full-process mechanization model for facility agriculture in desert and gobi regions is a complex and systematic project involving multiple technical fields. The aim is to provide a comprehensive mechanization solution for greenhouse agriculture in these areas, enhancing production efficiency, reducing labor costs, and achieving precise and sustainable agricultural production. A critical component of this research is the study of the yellow sand substrate cultivation model, which lays the foundation for the entire mechanization system. This model leverages the region's abundant sandy resources, creating a growing substrate suitable for crops through proper mixing and improvement. Researchers need to thoroughly explore the physical and chemical properties of the yellow sand substrate, optimize its water and nutrient retention capabilities, and develop a sustainable cultivation technology system tailored to different crops' physiological needs. During the land preparation phase, designing and testing a self-propelled ditching and mulching machine for yellow sand substrates is particularly important. This machine must adapt to the confined space and unique environment of greenhouses, performing tasks such as ditching, substrate laying, and mulching with precision. The design process should consider the machine's flexibility, operational accuracy, and environmental adaptability. For example, a modular design could allow the machine to be quickly adjusted according to different greenhouse layouts. Additionally, an intelligent control system should be developed for automatic navigation and precise operation to maximize work efficiency and quality. Testing should involve multiple trials in actual greenhouse environments to continuously optimize the machine's performance parameters and ensure it can handle various operating conditions.

Seedling sowing is a critical phase in desert and gobi facility agriculture, and research into intelligent identification and transplanting technology for tray seedlings in these regions presents a promising opportunity for automation. This research leverages cutting-edge advancements in computer vision, artificial intelligence, and precision mechanical control. The intelligent identification system can quickly and accurately evaluate the growth status, size, and quality of tray seedlings, providing valuable insights for the subsequent transplanting process. The transplanting technology must address the unique characteristics of the yellow sand substrate by developing mechanical devices that can precisely handle and plant seedlings. This technology significantly enhances seedling efficiency and ensures consistent transplanting quality, thereby establishing a strong foundation for effective crop management. Building on research into intelligent identification and transplanting technology, the design and testing of intelligent transplanting devices become pivotal in advancing mechanization in facility agriculture. These devices should integrate multiple functions, including automatic seedling extraction, precise positioning, flexible gripping, and orderly planting. The design must accommodate the unique greenhouse environment, considering factors such as space constraints, high temperature, and humidity. The device should demonstrate adaptability and reliability to handle various sizes and types of tray seedlings. Additionally, a user-friendly interface should be developed to enable operators to easily control and monitor the entire transplanting process. The testing phase should involve extended evaluations in real production environments to assess the device's efficiency, transplanting success rate, and impact on seedlings, with continuous design improvements based on feedback.

The selection and use of mechanized equipment are critical for advancing facility agriculture in desert and gobi regions. It is essential to choose machinery that caters to the specific needs of various crops and production stages, while also addressing the unique environmental challenges of these areas. Proper maintenance and management of the equipment are vital to ensure smooth operation and high efficiency. For soil treatment, deep plowing and rotary tillers can enhance soil structure, increase aeration, and improve water retention. During sowing and transplanting, precision seeders and transplanting machines can increase planting density and seedling survival rates. In the harvesting phase, efficient harvesters and grading machines can boost both the speed and quality of the harvest. Optimizing mechanized operation processes and technical standards is another key research area. Developing scientific operation procedures and guidelines aids farmers and technicians in completing various production stages more efficiently, thereby improving the overall mechanization level. Safe operation and regular maintenance of machinery are crucial to avoid production interruptions and efficiency losses due to equipment failures. Advanced management models and information technologies should be promoted to establish robust monitoring and management systems, facilitating intelligent and refined management of mechanized operations. Research on full-process mechanization in desert and gobi facility agriculture requires integrating findings from various aspects to create a comprehensive mechanization system. This system should encompass the entire process, from land preparation and seedling sowing to transplanting, water and nutrient management, and pest and disease control. Coordination between different stages is essential to ensure a smooth production flow, such as aligning the operational parameters of self-propelled ditching and mulching machines with the requirements of subsequent transplanting devices. Integrated water and nutrient systems should be designed to complement the needs of spraying machines and avoid interference. A unified intelligent control platform should be developed to enable information sharing and coordinated operation among various pieces of equipment. This platform should collect and analyze data from the entire production process to support informed management decisions. It is also important to explore ways to minimize mechanical interference with the greenhouse environment, such as controlling noise, reducing vibrations, and minimizing dust. For water and nutrient management and pest control, precision application and recycling are emphasized to reduce resource waste and environmental pollution. For example, developing a water and nutrient recycling system to treat and reuse wastewater and employing integrated pest management techniques that include biological control can decrease reliance on chemical pesticides. Research should also address the needs of different scales and types of facility agriculture. Highly automated systems can be developed for large commercial greenhouses to enable fully unmanned operation, while medium and small family farms require more flexible and user-friendly small-scale machinery. Additionally, equipment should be adaptable to specific crop varieties, with machinery designed to provide structural support for crops requiring trellises and specialized pruning and harvesting tools for fruit trees. Furthermore, research should explore how mechanization can enhance product quality, such as developing intelligent harvesting machinery that precisely controls picking times and methods to ensure the freshness and taste of agricultural products.

To promote mechanization technology in desert and Gobi regions, it is crucial to strengthen technical training and support services. Offering diverse, multi-level training programs will equip farmers and technicians with essential knowledge and skills for mechanized operations, thus enhancing their practical operation and management capabilities. Additionally, establishing a comprehensive technical service system is essential for providing timely and effective support, addressing various issues encountered during production.

In summary, researching and applying the full-process mechanization model for facility agriculture in desert and gobi regions accelerates agricultural modernization in these areas by enhancing production efficiency and quality while promoting sustainable regional economic development. This approach also provides a significant example and guide for the broader modernization of agriculture in China. Therefore, strengthening research and promotion of mechanized operations in these regions is crucial for advancing agricultural technological innovation and progress, and for achieving green, efficient, and sustainable agricultural development.

10.2 Full-Process Mechanization Technology Model

The full-process mechanization technology model for facility agriculture in desert and gobi regions encompasses every stage of production from land preparation to post-harvest processing. By integrating the specific environmental conditions of these areas with advanced agricultural machinery, an efficient and precise workflow is established. In the land preparation phase, targeted soil improvement is essential, starting with base fertilization using specialized machines to evenly distribute organic fertilizers and soil amendments, thereby enhancing soil fertility and water retention. This is followed by soil cultivation tasks such as deep plowing, rotary tilling, and bed forming. A self-propelled ditching and mulching machine designed for sandy soil performs these operations effectively. Seedling cultivation typically occurs in specialized greenhouses, where automated equipment, including seeders and irrigation systems, boosts efficiency and quality. For transplanting, an intelligent device with features like automatic depth adjustment and integrated water and nutrient systems is used, which minimizes soil compaction and reduces manual intervention. Field management focuses on irrigation, fertilization, and pest control. An intelligent water and nutrient system offers precise irrigation and fertilization, using soil moisture sensors and weather stations for real-time monitoring, and incorporating corrosion-resistant drip irrigation technology. Pest control is managed with an intelligent sprayer featuring image recognition for precise targeting and electrostatic spraying technology to improve efficiency and reduce drift. During harvesting, specialized machinery is selected based on crop type, with automatic harvesters for leafy vegetables and intelligent picking robots for fruit vegetables, designed to withstand the harsh desert and gobi conditions.

Through the full-process mechanization workflow outlined above, facility agriculture in desert and Gobi regions can achieve efficient and precise production. This model significantly enhances production efficiency and reduces labor intensity while ensuring effective resource utilization and environmental sustainability. Implementing this model requires accounting for the unique challenges of the desert and Gobi environments, such as high temperatures, strong winds, and dust, which necessitates selecting machinery capable of withstanding these conditions. Additionally, improving operator training is essential for effective equipment operation and maintenance. Establishing a comprehensive mechanization management system will enable real-time monitoring and data analysis of operations. Continuous technological innovation and equipment enhancement are necessary to further improve efficiency and precision. Integrating Internet of Things (IoT) and big data technologies will enable intelligent and digital management of agricultural production. Adopting this full-process mechanization model will propel facility agriculture in desert and Gobi regions into a new phase of development, significantly contributing to the modernization and sustainable development of agriculture in China (Fig. 10.1).

Fig. 10.1 Full-process mechanization technology model in facility agriculture

10.3 Key Equipment for Supporting Mechanization

10.3.1 Small-Scale Ditching and Mulching Machine for Yellow Sand Substrate in Greenhouses

Researchers need to delve into the physical and chemical properties of yellow sand substrates to optimize water and nutrient retention while aligning with the physiological needs of various crops. This will aid in developing a sustainable cultivation technology system. In the land preparation phase of desert and Gobi facility agriculture, designing and testing a self-propelled ditching and mulching machine for yellow sand substrates is crucial. This machine must perform precise operations in greenhouse environments, with flexibility, operational precision, and environmental adaptability being key considerations. Optimized ditching blade parameters include a cutting face height of 65 mm, a soil throwing angle of 30°, a width of 12 cm, and a lateral soil throwing width of 30 cm. Simulation analysis with EDEM software determined the optimal parameters: a working speed of 0.5 m/s, a blade shaft speed of 200 r/min, and a ditching depth of 30 cm. Field tests confirmed that the equipment met national standards with an average ditching depth of 31 cm, depth stability of 87.1%, ditch width of 41.3 cm, bottom width of 30.4 cm, and a soil fragmentation rate of 92.4%. This addressed traditional equipment issues and improved mechanization levels. Future research could further enhance blade materials for wear resistance and longevity, optimizing performance for sandy soils. Overall, this small-scale ditching and mulching machine significantly advances mechanization in facility agriculture, offering substantial practical value.

This research has resolved issues with traditional greenhouse equipment, such as poor ditching quality and suboptimal soil throwing performance, significantly advancing mechanization in facility agriculture in non-cultivated areas. It offers a

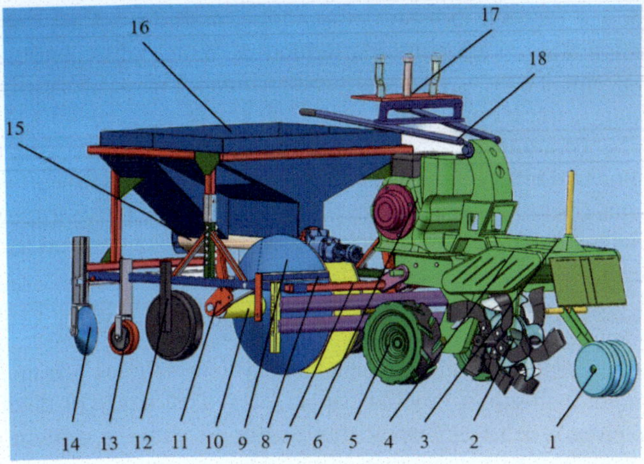

1. Guidance and Depth Control Device 2. Plow 3. Upper Soil Shield 4. Side Soil Shield 5. Drive Wheel 6. Engine 7. Frame 8. Film Hanging Frame 9. Compaction Device 10. Compaction Roller 11. Multi-Link Adjustment Mechanism 12. Walking Wheel 13. Film Pressing Wheel 14. Soil Cover Disk 15. Auger 16. Substrate Box 17. Gear Adjustment Lever 18. Control Lever

Fig. 10.2 Small-scale trenching and film mulching machine for solar greenhouse cultivation on a yellow sand substrate model

crucial reference for mechanized operations using yellow sand substrates in greenhouses. Future research should focus on optimizing blade materials to enhance wear resistance and lifespan, addressing the demands of high-frequency operations in sandy soils (Fig. 10.2).

10.3.2 Design and Testing of Intelligent Tray Seedling Transplanting Device

The design and testing of an intelligent tray seedling transplanting device for desert and Gobi regions aim to address specific challenges faced in facility agriculture in southern Xinjiang, an important agricultural area in China. This region experiences a dry climate, scarce winter precipitation, and low temperatures, which severely impact the efficiency and yield of traditional farming methods. To tackle these challenges, southern Xinjiang has actively promoted facility agriculture, using greenhouses, tunnels, and other structures to create suitable growing environments for crops. Facility agriculture offers several advantages over traditional open-field cultivation, including stable temperature and humidity for vegetables, protection from wind and sand, and reduced moisture evaporation, marking a shift toward mechanization and automation in agricultural production. With rapid economic and technological advancements, facility agriculture in China grew to 2.85 million hectares by 2022, making it the largest in the world. The area continues to expand, with facility tomatoes becoming increasingly important in southern Xinjiang due to their simple cultivation and high yield. However, current facility tomato cultivation relies mainly on manual operations, lacking mechanization, which limits improvements in production efficiency and quality and fails to meet the market demand for automated greenhouse vegetable cultivation (Fig. 10.3).

To address these issues, researchers have designed an intelligent tray seedling transplanting device tailored for desert and Gobi regions. The core component of this device is the seedling picking mechanism, which includes a frame, conveyor belt, stepper motor, ball screw, flexible grippers, and a camera. The device operates by using the conveyor belt to transport tray seedlings, with the camera identifying and locating them. The ball screw then drives the mechanical grippers to move to the designated position and grasp the tomato seedlings' stems.

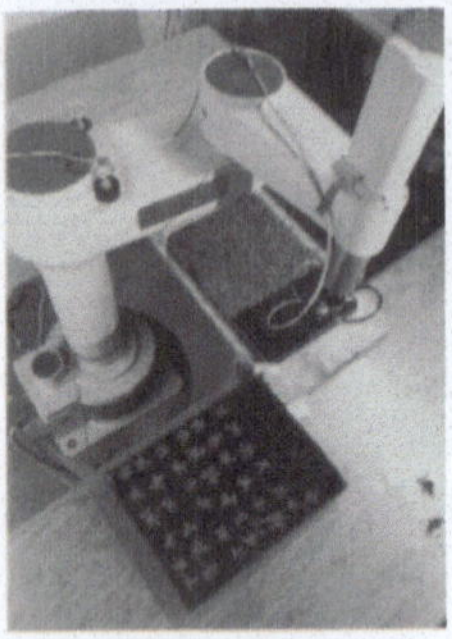

(a)SCARA-type transplanting machine diagram

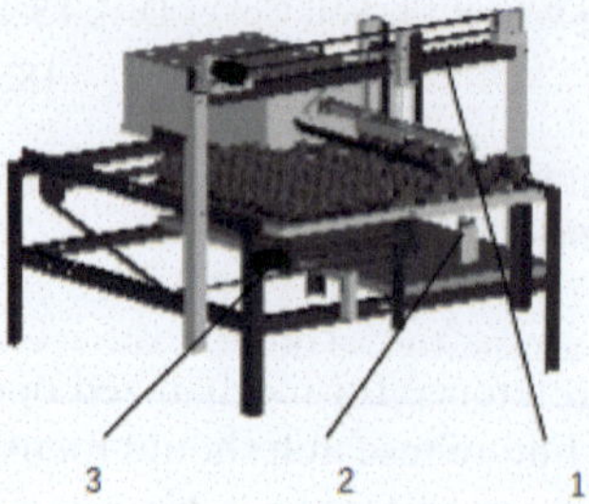

(b)Lettuce automatic transplanting machine

(c)Pic-O-Mat Vision transplanting machine

(d)PC-21 greenhouse plug tray automatic transplanting machine

Fig. 10.3 Design and experiment of an intelligent transplanter for plug seedlings

The design requirements for the seedling picking device include using cost-effective and durable materials, maintaining a simple structure, employing a stem-clamping seedling picking method, integrating depth camera recognition, and optimizing the non-interfering operation of various mechanisms. Key components are selected with precision; for example, the controller is an Arduino Uno R3 development board. The ball screw travel is chosen based on frame dimensions, with 700 mm for horizontal and 400 mm for vertical movements. For camera selection, the Intel RealSense D435i depth camera is used to obtain three-dimensional position information of the tomato seedling stems. This camera provides depth detection within a range of 0.2–10 m, with a maximum RGB resolution of 1920 × 1080, depth resolution of 1280 × 720, and a depth frame rate of 90 FPS, making it suitable for close-range positioning. The control system of the seedling picking device relies on the Arduino Uno development board as the core controller, programmed to control the motors and grippers for seedling picking. The system's circuit design involves Arduino controlling the stepper motor via a driver and directly managing servos and limit sensors. Software programming is done using Arduino IDE with C/C++ language. The entire seedling picking process is automated and includes camera recognition and positioning, mechanical gripper movement to the specified location, seedling grasping, and return to the origin. Limit sensors are employed to control the range of mechanical movements and prevent accidents. The specific workflow is as follows:

1. The camera identifies and locates the position of the tomato seedlings.
2. The mechanical gripper starts from the origin, with the horizontal ball screw driving it to the designated position.
3. The mechanical gripper opens, and the vertical ball screw moves downward to the front of the tomato stem.
4. The conveyor belt moves forward until it touches the limit switch, then moves forward by n-1 trays (where n is the current row of seedlings).
5. The mechanical gripper closes to grasp the tomato seedling.
6. The vertical ball screw moves upward to return to the origin.
7. The horizontal ball screw returns to the origin, while the conveyor belt moves backward by one tray distance.
8. Upon reaching the origin, the mechanical gripper opens to release the seedlings.

To validate the performance of the seedling picking device, researchers conducted a series of experiments. The results demonstrated high accuracy in identifying and grasping tomato seedlings, with an average success rate of 92.9% and a damage rate of 7.2% across 10 trials. These outcomes indicate that the device performs well overall. However, areas for improvement were identified. The primary factor affecting the picking success rate is the system's ability to recognize the condition of the tomato seedlings. When the system fails to identify seedlings with poor growth conditions, the mechanical gripper does not receive commands to pick these seedlings, thereby impacting the success rate. Additionally, lighting conditions significantly affect the recognition accuracy of the tomato stems, suggesting that the system's adaptability to different environments needs further enhancement (Fig. 10.4).

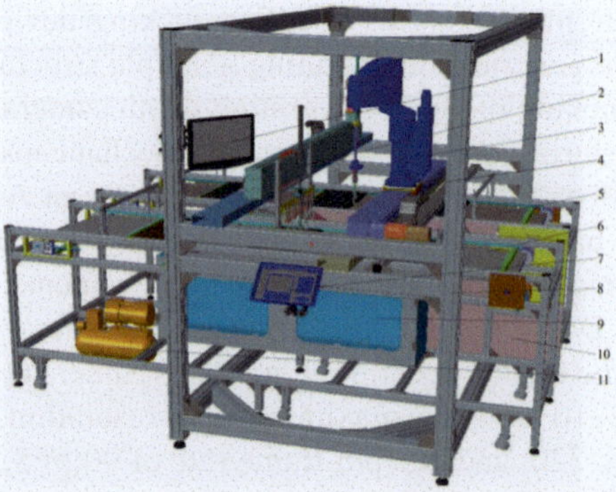

1. Display Screen 2. Multi-Claw Transplanting Mechanism 3. Single-Claw Replanting Mechanism 4. Detection Device 5. Target Tray Conveying Mechanism 6. Source Tray Conveying Mechanism 7. Touch Screen 8. Platform 9. Control Module 10. Seedling Discard Collection Device 11. Pneumatic Module

Fig. 10.4 Smart sorting-transplanting-replanting integrated machine for greenhouse plug tray seedlings

Overall, this seedling picking device offers significant technical reference for the intelligent development of facility agriculture in southern Xinjiang. It addresses issues of low efficiency and high labor intensity associated with manual seedling picking, thereby improving transplanting efficiency and quality. However, further research and optimization are needed in practical applications, focusing primarily on the following aspects:

1. Enhance the accuracy of the recognition system, especially under varying lighting conditions.
2. Optimize the design of the mechanical claws to further reduce damage to the tomato seedlings.
3. Improve the overall stability and reliability of the system for continuous long-term operation.
4. Explore the application of this technology to other types of seedling trays.
5. Research how to better integrate this device into existing facility agriculture production processes.

Seeding and seedling cultivation are critical stages in desert and Gobi facility agriculture, and research into intelligent identification and transplantation technology for seedling trays makes automation feasible. By integrating advances in computer vision, artificial intelligence, and precision mechanical control, intelligent recognition systems can quickly and accurately assess the growth status, size, and quality of seedling trays, providing valuable decision-making support for subsequent transplantation tasks. The transplantation technology must also address the characteristics of sandy substrates, developing mechanical devices that can precisely grasp and plant seedlings. This approach significantly enhances seeding efficiency and ensures consistency in transplantation quality, establishing a solid foundation for subsequent crop management. This research is not only crucial for the development of facility agriculture in southern Xinjiang but also offers valuable insights for the mechanization and automation of facility agriculture across other regions of China. As technology continues to advance and optimize, similar intelligent transplantation devices are expected to see broader application, driving facility agriculture in China toward greater efficiency and intelligence.

10.3.3 Design and Optimization of Smart Water and Fertilizer Integration System for Desert and Gobi Facility Agriculture

With global climate change and resource shortages worsening, agricultural production faces unprecedented challenges. In harsh environments like deserts and Gobi regions, efficient agricultural practices have become crucial. The vast deserts and Gobi regions of southern Xinjiang present opportunities for facility agriculture, but they also contend with water scarcity, severe soil salinization, and low fertilizer utilization. Facility agriculture, which can control environmental conditions and enhance resource efficiency, is being promoted in these areas. However, current water and fertilizer management models largely rely on experience and manual operations, resulting in low efficiency, resource wastage, and environmental pollution. To address these issues, the smart water and fertilizer integration system has emerged. This system monitors environmental parameters in real time and uses precise control algorithms to adjust water and fertilizer applications according to the crops' needs, ensuring efficient resource utilization.

In China, research on integrated water and fertilizer technology began relatively late. The technology's development started with the introduction of drip irrigation equipment in the 1970s, progressing through stages of early exploration, pilot promotion, and stable popularization. In 1980, China developed its first generation of complete drip irrigation equipment, marking the start of integrating water and fertilizer technology with sprinkler and drip systems. Extensive testing and demonstrations followed, leading to nationwide promotion. In 1988, Xu Zhiong systematically summarized the main fertilization methods, laying the groundwork for modern integrated water and fertilizer technology. By 1996, the Xinjiang Production and Construction Corps applied this technology to large-scale cotton cultivation. Around 2000, it entered a rapid promotion phase and was widely adopted in facility agriculture, significantly boosting crop yields. With the formation of China's water-soluble fertilizer industry and improved supporting products, a preliminary industrial chain for water and fertilizer integration equipment has developed. Driven by water-saving and reduced fertilization policies, integrated water and fertilizer technology has become increasingly popular across the country.

The unique geographical and climatic conditions of desert and gravel areas, such as ample sunlight and dry climate, offer potential advantages for facility agriculture. However, the main challenges in these regions are efficient water use and precise fertilization management. Traditional methods often result in low water and fertilizer utilization rates and significant resource wastage. The intelligent water and fertilizer integration system addresses these issues by incorporating IoT, sensor, and automatic control technologies. It monitors key parameters like soil moisture, electrical conductivity, and temperature and humidity in real-time, and automatically adjusts water and fertilizer applications based on crop needs. This system not only enhances water and fertilizer efficiency and reduces production costs but also improves the crop growing environment, leading to increased yield and quality.

The book designs an intelligent water and fertilizer integration system suitable for desert and gravel facility agriculture. The system mainly consists of the following components:

1. Water Source Engineering: Provides a stable irrigation water source, typically utilizing rivers, lakes, and groundwater.
2. Head Hub: The core component, including filtration devices, fertilization system, pressure and flow monitoring and protection systems, and automation control system. It pressurizes the water source, delivers filtered water to the fertilization system, and injects the mixed liquid into the field pipe network.
3. Fertilization System: Accurately injects fertilizer into the water flow to achieve precise fertilization.
4. Monitoring and Control System: Ensures the safe operation of the system and controls the fertilization and irrigation processes to meet crop needs with appropriate water and fertilizer ratios and application amounts.
5. Field Pipe Network and Irrigation Devices: Distributes the water-fertilizer mixture to the crop roots (Fig. 10.5).

The system control strategy employs a PLC-based approach and includes the following aspects:

1. Analog-to-Digital Conversion Program: Converts various sensor-captured analog signals into digital signals for system processing and analysis.
2. Operation Time Lock Program: Ensures the system operates within a set time period to prevent unnecessary water and fertilizer waste.
3. Automatic Irrigation Program: Controls irrigation start and stop based on soil moisture sensor data.
4. Automatic Fertilization Program: Controls fertilizer addition automatically according to the set electrical conductivity values.
5. Recirculating Irrigation and Fertilization Program: Facilitates the recirculation of clean water and fertilizer solutions to meet the needs of crops like strawberries that require frequent, light fertilization.

The Human–Machine Interface (HMI) design uses the Haiwell C7S-GW. Users can issue control commands and monitor equipment status through this interface. The design is simple and intuitive, making operation and monitoring easier (Fig. 10.6).

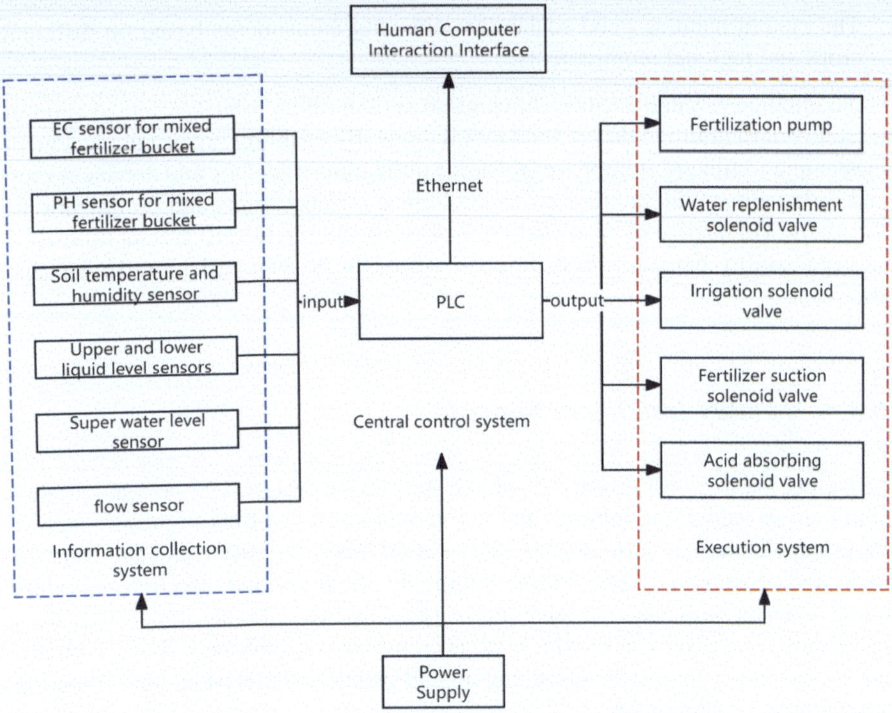

Fig. 10.5 Water-fertilizer integration machine control architecture diagram

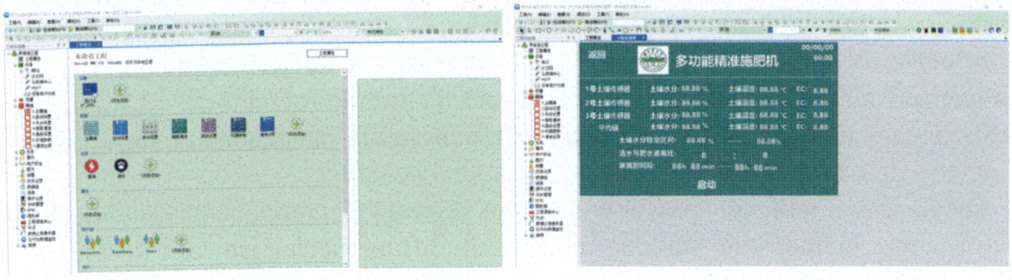

Fig. 10.6 Configuration software design interface

Innovations and Applications:

1. Developed a highly adaptable water and fertilizer distribution model tailored for strawberry cultivation in southern Xinjiang.
2. Utilized fuzzy PID control strategy to enhance system adaptability to environmental changes and improve control accuracy.
3. Achieved precise and automated water and fertilizer management, significantly increasing resource utilization efficiency.

4. The system features good scalability, allowing adjustments based on different crops and regional requirements.

The intelligent water-fertilizer integration system offers robust technical support for facility agriculture in desert and Gobi regions. By enabling precise management of water and fertilizers, it improves resource utilization efficiency and enhances crop yield and quality. With continuous technological advancements and optimizations, this system will be pivotal in advancing facility agriculture toward greater efficiency and sustainability, benefiting both desert areas and the broader agricultural landscape nationwide.

10.3.4 Facility Intelligent Sprayer

The facility intelligent sprayer is a sophisticated device designed for pest and disease control in protected agriculture, such as greenhouses. It offers high automation, uniform application, precise dosing, and reduced labor intensity compared to traditional manual spraying. This enhances spraying efficiency and effectiveness while cutting down on pesticide use and minimizing environmental pollution. Widely used in greenhouses and nurseries for high-value crops like vegetables and flowers, intelligent sprayers are continually advancing in intelligence and precision due to ongoing technological developments.

In recent years, China has made significant strides in developing facility intelligent sprayers. For instance, China Agricultural University has created a spraying robot that uses visual navigation technology for autonomous path planning and precise spraying. Jiangsu University has developed an intelligent sprayer combining crop image recognition with variable rate spraying for tailored application. These sprayers are already widely utilized in advanced agricultural regions of China, such as Shouguang in Shandong, Hangzhou in Zhejiang, and Changzhou in Jiangsu, where they have delivered notable economic and social benefits.

The facility intelligent sprayer consists of several key components. The mechanical structure includes the walking mechanism, lifting mechanism, and spray arms, which support and maneuver the equipment. The power system, usually driven by an electric motor, supplies the necessary energy for operation. The spraying system features the chemical tank, pump, and nozzles for storing, pressurizing, and applying pesticides. The control system integrates the controller, sensors, and actuators to manage and automate the spraying process. Finally, the human–machine interface provides operational control and status display, offering users an intuitive way to interact with and monitor the system.

The intelligent sprayer operates by first setting the spraying parameters, such as pesticide volume, pressure, and walking speed, based on the crop type and growth condition. Once activated, the sprayer moves along a pre-set path autonomously. It uses sensors to continuously monitor crop conditions and environmental parameters. The control system adjusts the spraying parameters dynamically based on this data

Fig. 10.7 Intelligent sprayer for facility agriculture

to ensure precise pesticide application. After completing a round of spraying, the device automatically returns to the starting point.

The intelligent sprayer is a key tool in modern facility agriculture, enhancing spraying efficiency, reducing pesticide use, protecting the environment, and ensuring operator health. With continuous technological advancements, these sprayers are expected to evolve towards greater intelligence, precision, and multifunctionality, contributing significantly to the sustainable development of facility agriculture. As costs decrease and performance improves, intelligent sprayers are anticipated to become widely adopted, becoming an essential component of facility agricultural production (Fig. 10.7).

10.4 Summary

The full mechanization model for vegetable production in solar greenhouses represents a major trend in modern agriculture, aimed at boosting efficiency and quality through advanced machinery and scientific management. This model reduces labor intensity and enhances land and resource utilization, supporting sustainable agriculture. Solar greenhouses, offering an ideal environment for year-round vegetable production, are crucial for achieving full mechanization. When designing and constructing these greenhouses, it is vital to consider mechanized operation needs, such as optimal space layout, adequate passage width, and appropriate slope. Additionally, modern temperature control, irrigation, and ventilation systems are essential to maintain the optimal conditions for vegetable growth.

The full mechanization operation procedures are central to this model, covering the entire vegetable production process from pre-planting to post-harvest. This includes fertilization with specialized machinery like fertilizer spreaders, soil preparation using deep and rotary tillers to enhance soil structure, and automated seedling equipment for improved seed utilization and quality. Automated transplanters increase efficiency and ensure consistent plant spacing, while smart fertigation systems and automated plant protection equipment manage irrigation, fertilization, and pest control precisely. Harvesting and transportation use equipment tailored to vegetable characteristics to boost efficiency and minimize damage, and straw shredders convert crop residues into organic fertilizer, supporting resource recycling. Key mechanized equipment includes deep tillers, rotary tillers, automated seedling machines, transplanters, smart fertigation systems, plant protection tools, and specialized harvesters and shredders.

Adopting the full mechanization model for vegetable production in solar greenhouses offers several significant advantages. Mechanized operations significantly enhance work efficiency and reduce labor input, enabling higher vegetable yields from the same land area. Precision management and standardized operations ensure consistent growing conditions, improving product quality and uniformity. While the initial investment in equipment is high, mechanization can substantially reduce long-term labor costs and resource consumption, improving economic returns. The precision in fertilization, irrigation, and pest control reduces chemical use and lowers environmental pollution risks. Mechanization also stimulates related industries, such as agricultural machinery manufacturing and smart control systems, driving upgrades across the entire agricultural supply chain. Modern production methods attract younger people to agriculture, addressing the issue of aging rural labor. Precision management optimizes water, fertilizers, and pesticide use, minimizing waste and achieving efficient resource utilization. Scientific management and resource recycling protect soil health, maintain ecological balance, and support sustainable agricultural development. However, promoting full mechanization in solar greenhouses presents challenges. The significant financial investment required for advanced machinery can impose economic pressure on small-scale farmers. Operating and maintaining modern machinery demands specialized knowledge and skills, raising competency requirements for farmers. Regional and crop-specific needs may necessitate tailored equipment, complicating the selection of suitable machinery. Additionally, full mechanization requires farmland standardization, which may be challenging to implement in areas with complex terrain or small-scale cultivation.

The full mechanization model for vegetable production in solar greenhouses represents a crucial advancement in modern agriculture. By implementing scientific planning, strategically allocating machinery, and optimizing production processes, this model significantly boosts both the efficiency and quality of vegetable production, leading to greener, more efficient, and sustainable agricultural practices. This approach not only enhances farmers' incomes but also plays a vital role in ensuring national food security and advancing agricultural modernization. With continuous technological advancements and increased efforts in promotion, this innovative production model is anticipated to see broader application, infusing new vitality into the transformation and upgrading of China's agriculture.

The manufacturer's authorised representative in the EU is Springer Nature Customer Service Centre GmbH, Europaplatz 3, 69115 Heidelberg, Germany. If you have any concerns regarding our products, please contact ProductSafety@springernature.com

Printed and bound by CPI Group (UK) Ltd, Croydon, CR0 4YY
01/10/2026
02225939-0002